JUN 0 7 2017

Flagland

The History of a Swamp Farm

Flagland
The History of a Swamp Farm

by
Martha Jean Hubbard Stewart

© 2009 by Martha Jean Hubbard Stewart

All rights reserved. This book may not be reproduced in whole or in part without written permission from the publisher.

Published by AKA:yoLa
Columbia, Missouri
www.akayola.com

Previously published photos used by permission of Missouri Ruralist Magazine; Farm Progress Companies.
www.farmprogress.com

This book was designed by Yolanda Ciolli using photos and vintage graphic elements from the Heath-Hubbard family.

Library of Congress
Flagland: history of a swamp farm / Martha Jean Hubbard Stewart
New Madrid County farm history
biography
Little River Drainage System
one room grade school
Southeast Missouri State University
University of Illinois, School of Agriculture
diversified farming

ISBN-13:   978-0-9842288-0-5
ISBN-10:   0-9842288-0-2

Table of Contents

Dedication	I
In Appreciation	III
A Summer Flood	VII
Family History	IX
Foreword	XI

Chapter 1. Birth of a Farm	1
Chapter 2. Under an Aladdin Lamp	9
Chapter 3. Farm History via Newspaper Clippings	25
Chapter 4. Exploring Friendships, Alma's Letters	33
Chapter 5. Matters of the Heart, Alma's Letters	81
Chapter 6. Until the Last Kick-off, Ralph's Letters	117
Chapter 7. Wrapped up in Farm Life	157
Chapter 8. Schools in a New Land	179
Chapter 9. Stretching for Space in the Old Farm House	195
Chapter 10. Gone — Barns, Fences, Gates, and Stiles	205
Epilogue by William D. Stewart	221
Appendix	227

Dedication

To my sisters, Mary, Ruth, and Alice, I offer my recollections of how our farm and your love have shaped my life.

Not always were my decisions wise ones. Once I talked Mary into following me through the large culvert near our home. We exited plastered with cobwebs on our new coats with velveteen collars. Mother just laughed. Perhaps she approved of our valor in swiping clean the metal culvert so that water could more swiftly enter the big ditch.

~ In Appreciation ~

Thanks to Frank Nickell, Ph.D. of Southeast Missouri State University for the insights into history and his gentle way of prodding me to focus on the way our lives have been shaped by farming.

Thanks to Fran Reynolds, Ed.D., Dale Oberer, Teresa Stewart, Angela Stewart, M.D., and Don Stewart, Ph.D. for their editing skills, and to my patient husband, Bill Stewart, who proofread my manuscript in bits and pieces and wrote the Epilogue.

Thanks to my grandchildren, who provided the driving impetus for this book.

Thanks to Duane Dailey, a writer for the *Missouri Ruralist,* for his journalistic eye in appreciating what farmers, past and present, have contributed to our lives and for his contagious enthusiasm for agricultural journalism.

Above all, thanks to Yolanda Ciolli, an artist turned publisher who paints with more than words.

Flagland

The History of a Swamp Farm

by
Martha Jean Hubbard Stewart

A Summer Flood

It rained last night,
And from a leaden sky all day
    it's poured.
Now my flowers lie
A tangled mass of leaves and
    broken blooms;
My garden's beat into the sand.
My fruit tree boughs, heavy
    laden
Are bare or broken.
A flood of muddy water covers
    all my land.
Tall corn that yesterday waved,
    a field of green
Tomorrow's sun will wilt and
    rob of color.
The cotton laden with its
    colored blooms
Will soon stand bare, undressed
    and useless.
And all of this because it rained
    last night,
And rained, and rained,
And now the floods have
    covered all our land.

— Alma E. Heath

Missouri Farm Bureau News
June 30, 1948

⁓ Heath – Hubbard ⁓
History

Paternal Grandparents	Maternal Grandparents
Fred Clark Hubbard	**Joseph Murphy Heath**
(May 24,1876 – January 20, 1930)	(November 7, 1887 – June 30, 1959)
m. August 7, 1905	*m. August 1, 1909*
Martha Carolyn Koehn	**Martha Ellen Sweet**
(July 2, 1880 – January 6, 1959)	(October 22, 1889 – November 16, 1978)

Ralph Charles Hubbard   **Alma Etta Heath**
(December 28, 1909 – July 6, 1983)   *m. June 4, 1933*   (November 14, 1910 – January 25, 1990)

Four Daughters, Their Children and Grandchildren

Martha Jean Hubbard   *m. June 16, 1957*   William Donald Stewart
(February 12, 1934)   (November 30, 1933)

    Donald Ralph Stewart   (February 20, 1960)
    Angela Ruth Stewart   (June 15, 1962)
       Justin Stewart Ehrhardt   (March 25, 1992)
       Elizabeth Jean Stewart-Ehrhardt   (May 20, 1994)
    Teresa Jean Stewart   (September 19, 1965)
       Luke Evan Stewart-Jones   (January 12, 1995)
       Eleanor Grace Stewart-Jones   (March 10, 1998)
       Zora Hope Stewart-Jones   (September 19, 2002)

Mary Ellen Hubbard   *m. Nov. 16, 1966 (div. 1971)*   John Charles Jones
(February 12, 1934 – December 19, 2005)   (April 12, 1947)

    Mark Hubbard Jones (May 13, 1968)
       Kelli Marie Jones (July 11, 1990)
       Dakota Lee Jones (June 23, 1993)

Alma Ruth Hubbard   *m. July 30, 1961*   David Wayne Richardson
(June 16, 1939)   (January 9, 1932 – June 29, 1998)

    Joseph David Richardson   (December 16, 1965)
       Hannah Ruth Richardson   (September 24, 2002)
       John Zachary Richardson   (September 6, 2005)
    Stephen Wayne Richardson   (February 25, 1969)
       Annie Catherine Richardson   (August 15, 1997)
       April Hope Richardson   (April 30, 2001)

Marjorie Alice Hubbard   *m. December 28, 1968*   John Daniel Bourzikas
(June 16, 1945)   (August 3, 1946)

    Paul Daniel Bourzikas   (October 16, 1972)
    Grant Andrew Bourzikas   (July 13, 1976)
    Kent Matthew Bourzikas   (May 8, 1982)

～ Foreword by ～
Martha Jean Hubbard

During harvest season my husband, Bill, and I took our grandchildren, Justin and Beth, to the family farm in New Madrid County. We rode the combine in a field of rice, rambled down farm roads, captured woolly worms in a tin can, ate pears in a wagon, prowled through junk in the corn crib, and trudged through a deep ditch to the site of the one-room school I attended. I have tried to paint a picture with words of the bustling farm I knew. How does one compare country life to city life?

"I can eat these pears while I play house!"
–Beth Ehrhardt, age 5.

"Oh, Neanie," Justin finally exploded, "you mean I can run anywhere on this farm and explore everything by myself?"

"Yes, that is just what I mean," I answered, realizing he had caught a glimpse of the freedom I experienced daily on a farm. It offered a freedom to learn about life from first hand experiences. It opened doors to people and allowed me to see how adults and children coped with life's hardships and conquered the many tasks that farming demanded.

"A big stick is perfect for exploring a ditch."
– Justin Ehrhardt, Age 7.

This book started out as an attempt to tell my grandchildren about my life on a farm, but it is more convoluted than I thought. The complexities of life are caught in the stream of events that have taken place in our lives, in shaping the land, in trial and error, and in the multiplicity of changes in farming practices. This account spans births, deaths, struggles, illnesses, and celebrations, and yet the land continues to be.

My search through farm records offered up a treasure trove. I found a farm journal that Dad kept, documenting his farming practices and giving a thoughtful analysis of what is required to make a farm better. I found letters documenting a woman's life on a farm that Mother had written over a period of thirty-six years in a Round Robin format. Each woman in the Round Robin circle lived on a farm somewhere in the United States and added her letter to a stack that passed around the group every few weeks. In the days before television, the letters were a source of both entertainment and connection. For me, however, they documented the momentous events in our lives as well as the repetitive tasks of everyday farming. More recently, I found Round Robin letters that Dad had written to his four brothers describing his farming tasks, laced with his enthusiasm for sports. In dredging up the past, I looked anew at how the face of the family farm had changed.

Around the turn of the 20th century, land in the swamps of Missouri offered an opportunity for people of vision. Their dreams took place first on paper that set up a plan for moving dirt and funneling water into new channels to drain the swamps. Families moved in to take part in the reclamation project called "The Little River Drainage System." Sawmills were set up in the midst of trees, and land was cleared. The lowest land, called flagland after the flags of the swamp grass, was appropriated as desirable because there were no trees. Our farm includes some of this land. I am from one of those families who took on the task of shaping a farm out of the swamps. This is the story of the farm that I grew up on in the heart of New Madrid County, Missouri.

It was an awesome task confronting nature with muscles– the muscles of workers and mules. Chains were fastened around the stumps of trees, and mules struggled to pull them out of the mud. Plows and mules began to shape the dirt into fields, filling in the pits left by the stumps and digging ditches to remove the water from the land. As a child I watched my dad use a mule to plow the garden by our home and disk it, turning over the dirt, called gumbo, and breaking up the dense clods into a workable soil. I watched workmen take a team of mules and pull a wagon of manure from the barns. They shoveled out the contents onto the land to give the gumbo elasticity and fertility. I saw mules pulling a plow held in the ground by the muscles of a man. I peered through the wire fence at the lug tractors with their spiked metal prongs and watched them take on the task of working in the mud where rubber tires would spin. I saw the parade of bulldozers, scrapers, graders and ditchers take on the task of moving dirt into the lowest areas of a field. Finally, I watched the massive dirt movers push down the levees and sculpt the land into graded fields. Irrigation wells were dug, and fields were put to grade so they could be irrigated. How much of this do I tell my grandchildren?

Today, at the beginning of the 21st Century, we drove through the farm. The land had been burned to get rid of red rice and sheath blight. Portions had been re-graded to have straight levees for irrigation, and the fields had been disked for the new crops. The vast expanse of graded fields, free of barns, trees, stubble and devoid of even clods, made me misty-eyed. I recalled the patchwork of small fields tantalizing us with opportunities to climb gates, scrounge in barns, and explore the overgrown fence rows. Yet, I remember it is the same dirt that crumbled under the hands of my father and his father. I want my children and their children to know the farm that has shaped my life. May you, too, share my gumbo.

Chapter 1.
The Birth of a Farm

"Come quick! Mark is stuck in the mud in the garden. He's crying and can't move. Help us!" The appeals interrupted a friendly Sunday visit that I, Martha Jean, was having with my mother and sisters, Mary and Ruth. We promptly went outside and lined up by the garden's edge. There was four-year-old Mark in a pair of red rubber boots embedded in the plowed garden. The more he struggled to move, the deeper he sank. Cousins Teresa, Joe and Stephen hopped about and commiserated with Mark while the older cousins Don and Angel (Angela) spluttered orders, "Pull hard! Walk out!" Mark had given up. Frustrated, he stood with head bowed and refused to respond to anyone's pleas. We surveyed the water-soaked gumbo and opted for a verbal solution. "Wade out barefooted! Pull your feet out of the boots, then pull your boots out of the gumbo and bring them with you. We will clean the mud off you later."

Such was life on a gumbo farm. Our garage was always littered with boots encased in balls of mud. A bootjack was by the back door to assist in the removal of the slimy, caked boots. One did not venture out on a rainy day without boots. I delighted in the spongy, slick dirt that packed as hard as concrete when dry to offer a smooth surface for biking or running barefoot. Even more wonderful was the gumbo when it was the wet, sticky, soapy stuff that could be sculpted into clay figures or dishes for a playhouse. It provided us with an endless supply of Play Dough. We delighted in traipsing through the yard after a rain and knocking down the crawdad chimneys that sprouted up over the lawn like mushrooms, built by the mud-dabbing creatures from the ditches. We probed the holes of the chimneys with sticks but never found any crawdads.

The soil was black and rich. The sediment deposited by the Mississippi River grew rank weeds as well as productive crops without fertilizer. Technically it was "sharkey clay," but we referred to it as gumbo. (Soil Survey of New Madrid County, Missouri, by United States Department of Agriculture Soil Conservation Service, 1968-72) In dry weather, large cracks left plenty of space for crickets, spiders, bugs or crawdads. So large were the cracks that we could put a hand in and reach as far as an elbow. Oh, the problems in the rainy season! Many times I, like Mark, stepped out of my boots to regain a footing in the mud that was not unlike quicksand. It sucked my feet deeper into the ooze and closed around me like concrete. Our cars would slide sideways through wet places in the road, and make deep ruts in the black paste. At times, we would fly through the sucking mud at high speed, fearing to stop until we reached a firm footing. From such experiences we learned to manage the sticky stuff called gumbo.

Our farm was born out of the labor of the Little River Drainage District engineers. Land abstracts show that farmland went on sale in 1908, and would-be purchasers jumped at the opportunity to farm new land wrenched out of the swamps of Southeast Missouri. Among those purchasers were Fred and Martha Hubbard of Urbana, Illinois. The two had a dairy and fruit farm at Urbana and were

Flagland was an intriguing farm purchase for Fred and Martha Hubbard.

intrigued by the real estate agents who promoted the Missouri swampland. A soil analysis compared the land with the rich prairie soil Fred knew so well in Central Illinois. Many people at Urbana, particularly friends in the Agriculture Department at the University of Illinois, became interested and bought land in New Madrid County. The Hubbards initially bought 479 acres. Fred selected "flagland," or land without trees, because he thought it would save money in clearing the land. Unfortunately, it meant that the land was some of the lowest in the county and sat as a pond most of the time.

The battle to control the waters of the swamp and fashion the contentious gumbo into workable land began. It was a part of the reclamation project known as the Little River Drainage District. The overall plan called for north-south ditches to be dug every mile across the swamp, moving the water to a lake in Arkansas. Our farm house was built near the levee of one of the big ditches. Tenant houses and barns tended to be sprinkled along the banks of the levees of the big ditches. The first bridges were spindle-legged affairs across the twelve-foot deep ditch. Two lapping board planks spanned the water, resting on poles sunk into the bottom of the clay ditch. Mothers inched carefully across the bridges to their houses behind their children, straddling the boards. All were grateful indeed, when a handrail was attached to the bridge. By our house, a large bridge heavy enough to support a wagon and team of mules was built from the abundant cypress trees that did not rot. A concrete footing supported the heavy cypress timbers that were sunk in the gumbo. Above the supportive scaffold were added cross timbers with runways laid on top for the wheels of vehicles. Side boards were added to the edge of the bridge to catch any wheels that missed the treads. People who were not particularly courageous parked their vehicles beside the ditch and walked across the wooden bridge.

As a child, I stood at the fence in our front yard to watch the muddy water rush down the big ditch in front of our house. The trash and debris caught in the eddies of water were a source of delight as they spun around from the big culverts that drained

"Looks like a hip-boot farm to me." says Ralph Hubbard.

the fields. After a six-inch rain, the fields were virtual lakes. Smaller ditches in the fields filtered the water into the large ditch. Culverts under the roads drained the fields that stood inundated. It was a part of the vision of the engineers of the Little River Drainage District to drain the rich land of "Swampeast" Missouri through an elaborate network of drainage ditches, canals, and levees. It is interesting to note that a greater amount of earth was moved during the construction of the Little River Drainage District than was moved during the construction of the Panama Canal. (The Little River Drainage District of Southeast Missouri, a booklet authorized by the Little River Drainage District Board of Supervisors in 1989, p. 7.) The area was likened to the Nile Valley of Egypt, but to me it was the farm I lived on in the middle of New Madrid County. It was my home. Every rain became a battle of shovels, enlarging the smaller ditches, digging new ditches in the fields to drain the water off crops, and opening additional channels to the larger ditches. During dry times, ditches were re-dug, removing the silt that had built up. Periodically we sounded an alarm, "The dredge boat is coming! The dredge boat is coming!" We watched with fascination as the main ditch was re-dredged, throwing up a pile of gumbo that left the ditch deep and ready for a new deluge of water and provided the hump of dirt for a road that paralleled the main ditch.

My mother told the story of walking around a pond in the middle of the farm to go to school and being frightened because it was a pond in which a horse was reputed to have drowned. In the plowed fields we delighted in looking for and picking up what we called "yonkey pins," the seed of the flags or tall swamp grass that grew in the ponds. The scientific name is Nelumbo lutea, the American Lotus and was an important food plant for the Native Americans. The young shoots were eaten as vegetables, and the seeds were hulled and roasted. (Denison, Edgar. Missouri Wildflowers; Missouri Department of Conservation, 3rd edition, 1978) We cracked the seeds with a hard rock and ate the nutmeat inside. The taste was on the bitter side, but we felt akin to the Indians. Additional blocks of land were added to the original purchase until there were 969 acres in the tract of land on Catron Ditch.

From 1908 to 1918, Fred Hubbard visited the Missouri farm for periods as long as six weeks at a time and tried to encourage farming by two families, the LeSieurs and the Raineys. Sometimes my father and his brothers came with Fred Hubbard to the farm. He reported, "We thought we were on a campout—no running water, no plumbing, and no electricity. It was wonderful." A sawmill was located in the area, and clearing the surrounding land was an on-going operation. In 1918, a farm supervisor, Richard Trimble, was hired to take care of the farm. He moved with his family from Urbana, Illinois, to the Southeast Missouri farm. It was the only cultivated land in the area.

About this time my maternal grandparents, Joseph and Mattie Heath, came to the community with my mother, Alma Etta, who was eight years old. They saw an opportunity to own a farm at a reasonable price, and moved from Wayne County to New Madrid County, Missouri. Joseph worked in the sawmill and waited for the chance to farm. The land was far more undeveloped than he had anticipated. The houses were unfinished. The gumbo road, when muddy, was almost impassible. The water was red with iron, and the school was two and a half miles

The dream of purchasing a farm prompted the Heath's move to New Madrid County.

away. In 1921, Joe Heath gave up his dreams of farming and became the manager of the Hubbard farm. It was not until 1934 that he realized his dream of owning a farm by purchasing land that abutted the Hubbard farm for eight dollars an acre. Those were the days of the Great Depression.

Some speculation should be made about the New Madrid Earthquakes of 1811 and 1812. We lived with frequent earthquake activity. Dishes would rattle in the cupboards. Glasses would clink and dance on the shelves. We would stop, wait a moment, and comment, "Oh, we're having an earthquake today." As quickly as the quake had started, it subsided. As a child, I heard tales of the day the Mississippi River flowed backwards, and the land opened up and engulfed trees. This lifting up of the ground as the waves in the water moved upstream gave the illusion that the river was, in fact, flowing backwards. Stories that entire islands in the river disappeared embellished the colorful accounts. Every time we went to the town of New Madrid to shop, the entertaining sport was to go up the levee and look at "The River." The massive amount of water that poured down from the north took our breath away, and we stood quietly and watched the volume of water tear at the shore and hoist logs and debris on the long journey to the Gulf of Mexico. Family outings involved taking the Tiptonville Ferry across the river to Reel Foot Lake in Tennessee to picnic and swim. We would rent a boat and go out among the trees that had died and were left standing like so many toothpicks in the shallow lake. This was the lake that was formed in the cataclysmic upheaval of the New Madrid Earthquake of 1812.

At other times I would listen to my father speculate that the sandy places on the farm were perhaps a result of the quake. The land that undulated in waves and left fissures large enough to swallow trees or houses would leave a streak of sand among the gumbo. Or more likely, the river pouring into subterranean chambers would create a whirlpool effect known as a sand blow. Such sand blows are found in places on the farm. When the farm was graded for irrigation in the nineties, the workmen were amazed at the random sandy places in the pure sharky clay or gumbo. According to my father, the top soil was only four or five inches deep, typical of swamp land or land produced by trees. There was enough drainage to permit farming, however, and the soil was exceedingly fertile. The great swamp, beginning two or three miles below Cape Girardeau, Missouri, had once been the bed of a river, either the Mississippi or the St. Francis augmented by the waters of Castor River and Little River. It is interesting that my grandfather, Joe Heath, suggested that the "horse pond" in the middle of the farm was a part of the old Castor River bed. When the land sank, we do not know. We do know that the courses of rivers and streams were changed in a land poorly drained. Cypress trees were abundant on the farm, but there were also pecan, oak, ash, and sycamore which are common to higher and drier ground. A magnificent cypress tree in the front yard of the farm home provides an abundance of cypress balls for friendly fights in much the manner as a snowball fight. We would crawl on the roots that protruded around such trees, the knobby cypress knees. It was a land that had been reclaimed from a swamp.

Chapter 2.
Under an Aladdin Lamp

When Ralph C. Hubbard, son of Fred and Martha Hubbard of Urbana, Illinois, graduated from the School of Agriculture at the University of Illinois in 1931, it was during the depression. Land in Missouri was mortgaged against the land in Illinois, prime land near the campus of the University of Illinois. An offer to settle the mortgage on the Missouri land for ten dollars an acre was made, but no money was available. To save the Missouri investment as well as the Illinois land, Ralph's parents gave him an old Model "T" Ford and sent him to Missouri "to see what could be done." Plans called for Ralph, the new college graduate, to put his agriculture training into action with Joe Heath, as foreman of the Hubbard farm.

Complications developed. The Heath's daughter Alma was going to school at Southeast Missouri State Teachers College and had a job teaching at a rural school outside of Parma, Missouri. She stayed with a school board family during the week but came home on weekends. She was most vexed at having one of the Hubbard boys move in with them. Both Alma and Ralph were betrothed to other persons and had no intention of changing allegiance. But change they did! Engagements were broken. Alma saw in the vexing Mr. Hubbard a winsome

Being stuck on a gumbo farm prompted romance between Ralph and Alma.

idealism and a work ethic to turn the swamp land into a show farm following the diversification touted by the School of Agriculture at the University of Illinois. Ralph saw in the dark-haired lass Alma, a beauty that was reflected in the love poetry she taught her students and the gift she had of understanding people of all ages. It turned out to be a happy combination for all. Ralph Hubbard had the book learning, and Joe Heath had the experience and practical know-how. Joe could repair or improvise any equipment in the blacksmith shop. He was a natural veterinarian and did all the castration of farm animals. Although Joe went to work at twelve years of age at a saw mill and had no chance for an education, he had a natural eye for architecture and could build anything. It was exciting to him to listen to what Ralph had learned in school.

Alma Heath and Ralph Hubbard were thrown together at a propitious time, and so we have the beginning of a love story that lasted over fifty years. Their love knew no bounds. It was fed on their common interests, their grasp of ideals, and always their concern for others. In the love they held for each other, they found the strength to confront the problems that came their way and give solace to their family and the people in their community.

Ralph kept a journal of his farm operation. "The Journal" gives the reader an insight into how farming evolved into the sophisticated practices of the 21st century. He moved from a hand labor and mule operation to one that depended on mechanical tools. To build the soil, he integrated a hog and cattle operation along with tillable crops and pastures. His methods were molded by his youthful enthusiasm and idealism. We read in his journal his reflections and what he has learned. Ralph's circadian rhythm was set on "Early to bed and early to rise," so he would have sorted through his thoughts in the wee hours of the morning under an Aladdin lamp before the days of electricity. As a child I would get up and share a bowl of oatmeal with Dad at four a.m. The farm today is the result of what he established in his pattern to build up the fertility of the soil through diversity of farming practices. Ralph called it his "Farm Business

Notebook." It was conveniently indexed so that he could add conclusions year by year. Only excerpts have been provided.

The college graduate is prepared for the challenge of the Great Depression.

Ralph Hubbard's "Farm Journal"

It is my intention to record in this book my own experiences, thoughts, and opinions regarding farming in Southeast Missouri. While attending college it was always my belief that by so doing, I would later avoid many of the rather bitter lessons of experience. Now with four years of experience as a farmer, I intend to try to profit by the experiences I have had and if possible not make the same mistakes twice. The plan of this book will follow closely the index plan of the "Illinois Farm Account Book."

Pioneer farming was a job, not a business. Modern farming is a business with large investments on which fair profits must be made if the business is to be satisfactory. It is my opinion that New Madrid County in Southeast Missouri is a place where farming is still in the pioneer stage. A farmer should try practices recommended by Agriculture Experiment Stations. When I first started out, I certainly could not have afforded to try many things which failed. For that reason, I try to study carefully any practice I intend to try, then not try it on a large scale until it has proven itself.

    I cannot forget the time I got my fingers burned by buying a bushel of Morse Rose Cotton seed. My wife and I went to Memphis to the Mid-South Fair in 1933. While there we saw displays of Morse Rose Cotton which was advertised as the four-bale-per-acre cotton of exceptionally long staple bringing a premium, so I tried some. It certainly was a failure. I should have talked with the county agent or some real authority, and I would not have lost my eight dollars.

    <u>Drainage:</u> Probably no one factor holds this section of Missouri back worse than drainage. The most successful type of drainage here, undoubtedly, is the open ditch system. In the gumbo parts of this farm, tiles would seal over, and in sandy areas they would silt full. For making open ditches there are several different methods which have their particular adaptations. This spring I used a large Caterpillar tractor and grader, which belonged to the Delta Reality Company at Catron, for two ditches. It cost five dollars an hour to use this outfit, but I believe that it is the cheapest way of making large ditches. This outfit made a ditch nearly a yard deep and nearly a quarter of a mile long in an hour. Such work must be done, however, when it is dry. A light tractor such as I use for farm work and my old road grader does good work for shallow ditches and to re-clean some of the larger ones. A Martin Ditcher [a large lister] with two or four mules is all right for small garden ditches and may be used during a wet time. A lister [a double moldboard plow which heaps the earth on both sides of a furrow] may drain many a small pond during a wet time. However, it is mighty hard on mules. The best drainage work that I can do during a wet time is to draw maps and prepare for the work when it is dry. Drag lines (with heavy sledges or sleds) probably are the best method for ditches too large for the large Caterpillar and grader.

Fertility: No farmer can afford to farm in such a way as not to maintain fertility. I would like to have a productive farm, and for that reason one-fourth of this farm or more will be in legumes. It is a custom among farmers in New Madrid County to burn all crop residues such as corn stalks, crab grass, and cotton stalks. For a soil lacking organic matter as our soil here is, this practice certainly must be stopped if we are to maintain fertility. It is my experience in observing soil that the dark layer of soil is seldom over five or six inches thick. Soils of timber origin lack organic matter. This is one reason that much of our new ground, which for a few years after clearing produces high yields of corn, but soon produces less than half what it did when it first was under cultivation. Our soil is very much benefited by the use of legumes, and such a practice as turning under a crop of beans certainly means greatly the production of a crop of corn the following year.

Rotation: As you may notice, I am following four rotations. Fields for each year in the rotation are from 140 to 150 acres in size. A rotation of wheat-lespedeza [a shrubby plant used for forage] is planted each year. About half of this acreage is pastured with beef cattle. Sometimes oats are substituted for part of the wheat when weather conditions prevent seeding all of the wheat in the fall. If oats are seeded, they are used for pasture. In this five-year plan, the cotton follows two years of lespedeza. This saves more than half of the expense of controlling the weeds in the cotton because close pasturing and clipping have prevented the most troublesome cotton-weed seeds to form. Vetch [a trailing plant of the legume family] is seeded in cotton middles to be turned under for corn the following year. I have observed that corn grows well following the cotton on lespedeza land in this rotation. Beans are substituted for part of the cotton in this rotation when bean prices are high and labor scarce. I wait till the sixth of October to insure a stand of lespedeza next year and then merely use a heavy disc or field cultivator to prepare the seedbed on the lespedeza sod.

  I am planning a spring pit rotation on the west side of Catron ditch. This plan calls for four fields of twelve acres each. The rotation will be one year of corn-soybeans with three years of alfalfa. The pigs will be on the alfalfa the third year. I seeded one of these fields to alfalfa this spring using one-half bushel of oats per acre as a nurse crop, seeded with the alfalfa about the middle of March. I like the oats for spring-seeded alfalfa since the alfalfa does not drown as badly or fill with weeds as quickly, but do not want more than one-half bushel rate of seeding because it smothers the alfalfa too much.
  For the fall pigs I have another rotation on the east side of Catron ditch involving fields J, L, M & N of approximately ten acres each. This is also a four-year rotation of corn, soybeans, wheat (lespedeza), rye seeded in August with mixed clover grass. This mixed clover pasture stands over for the fourth year in the rotation on which the fall pigs are raised.

  <u>*Improvements:*</u> *While I believe no farmer should put many improvements on a mortgaged farm, he certainly should be ready to repair buildings. I have found that a dollar spent for repairs in time is worth more than any dollar anyone can spend for other improvements. I like to think of the old proverb, "A stitch in time saves nine."*

  <u>*Electric Fence:*</u> *I tried my first electric fence this week. The power used was a six-volt battery and a coil from a "T" Model Ford. It was very effective. I am likely to use it a lot in the future for temporary fences.*
  After using electric fences for eight years, I still find more and more use for them. It takes very little time to put them up if you have steel posts with straight cross pieces that you can place your foot on and just push in the ground.

  <u>*Livestock Farming:*</u> *I believe that livestock on a farm insure[s] profits; they are one of my insurance policies. They are one way to diversify. In 1937 crop prices were low, but I fed out some steers which brought $13.10 per cwt. [hundredweight]*

Farm Power: Mules are usually my cheapest source of farm power. My tractors do not go out until the mules are in the fields and not unless the mules do not provide enough power to keep the work up. Mules will break ground much cheaper than tractors in early spring. When hot weather arrives, mules cannot do nearly as much work as in early spring. In Southeast Missouri, there are usually many places in the fields where tractors bog down. When one takes a tractor to the field when it is likely to hang up, a tractor just cannot be used effectively. On the first of January, I saw a farmer operating his tractor on Sunday even though he had six teams in the fields. The farmer was really working his ground too wet. Land broken with the tractor cost two or three times as much as the mule breaking.

In 1948 my own picture of the relation between mules and tractors has changed with the high prices for labor which exist at present. In the period between 1932-39, I seldom used a tractor when mules were in the barns. Now I use the tractors first and the mules last when good farming conditions exist.

Costs of keeping Mules: We are feeding about 75 bushel of corn and three tons of alfalfa hay per mule, per year. We feed hay alone six months in the year. Figuring the hay at ten dollars per ton, and the corn at fifty cents per bushel, if you feed twice as much hay during the hay and corn period, it will cost less than half as much per month to feed the mules during the hay alone period.

Breeding Stock: There is no question but that a great many factors enter into the raising of large litters up to weaning time. The first matter to consider of course is the matter of breeding stock. Sows out of large litters are more likely to raise large litters. In general it has been my policy to discard a sow any time it raises less than six pigs. I have never seen sows that took care of pigs as well as the stock that was on the farm here when I came in 1931. We had sixty gilts and twenty old sows. At one time in the spring of that year we had 570 pigs which as you can *see was over an average of seven pigs for eighty brood sows.*

I feel that even though I have brought in some of the finest pure-bred breeding stock, my original grade stock was superior. I may go back to the original cross of Tamworths and Durocs because of the wonderful pig-raising abilities of the Tamworths. It seemed that as long as we had enough Tamworth blood to have an erect ear and not too blocky a sow, we had wonderful pig raisers, but with enough Duroc blood to droop the ear, it was harder to raise the pigs. The sows were lazy and not as inclined to fight. Also instead of carefully lying down on their stomach, they were inclined to flop down and crush pigs.

<u>Buying Feeder Pigs:</u> I had planned to have sixty litters of pigs during August and September of 1937 but due to a combination of factors succeeded in raising only about a hundred pigs. With a good corn crop and the price of corn low and hogs high, I decided to purchase some feeders. Accordingly, I purchased some two hundred of a rather assorted lot from different sources. From my experience with these I would like to point out some of the pitfalls to look out for in the purchase of feeders. (1) The first and most important is the matter of prices. I have been watching the hog market very carefully during the last six or seven years, and from my experience I have observed the low is generally around Thanksgiving time. (2) Buy a feeder that will weigh 240# by March first. Don't buy pigs weighing over 75 pounds.

<u>Advantages of Fall Pigs:</u> My plan of raising calls for the production of both spring and fall litters of pigs. There are several reasons why I plan to have most of the pigs farrowed in August and September. The most important reason is that pigs born at this time can be farrowed on clean ground. They have no mud handicap, and I get an early turnover of the corn crop. August and September pigs require less housing problems, shade being the most important item. They also are large enough by cold weather to continue good growth, and by the time muddy conditions force them to higher feed grounds or old lots, they are beyond four months of age when worms bother them less. Pigs

born in warm weather are less likely to be crushed by the sows since they will not huddle close to the sow for warmth as in the winter. Another reason why I plan a heavy fall pig crop is that it

Movable A-frame hog houses made farrowing pigs on clean ground feasible.

provides more elasticity. During years of heavy corn crops, one may raise a large spring crop of pigs and feed all this crop; whereas if he planned on few fall pigs, it would be harder to feed all his corn. This early turnover means getting the corn marketed before others can raise a lot of hogs and depress the market.

<u>Crossbred Hogs:</u> After eight years of raising crossbred hogs I am inclined to believe that crossbreeding of purebred hogs, a practice for a number of years recommended by the Minnesota and Missouri stations, has little if any merit. There have been times when it seemed very successful and other times when it certainly did not produce better results than any ordinary hogs. The main disadvantage to the system lies in the fact that your herd of hogs may look very much like a herd of scrubs if sickness occurs in your herd. You tend to have hogs of vari-

ous colors, shades, and types. It is my belief if we are to make progress with hog breeding, we will be more likely to get results by the production registry method. From now on it is my intention to breed production into purebreds rather than to try crossbreeding.

Bees: While I raised bees the first few years I was in Southeast Missouri, in recent years I have more or less lost interest in them although I believe they could be very profitable. The same factors which give us a month longer pasture season and more mild winters than over much of the state would make bees do well here.

Corn: We raise fair corn; we do not raise corn as well as farther north. In 1955 the yield was about 30 bushels on heavy land and about 50 on mixed land with negligible difference between open pollinated and the two hybrids. Heavy land was too wet for the first eight weeks, and then so dry it cracked open the last eight weeks.

Wheat: Wheat is a gamble because of rainy winters. I consider wheat an incidental crop in the production of lespedeza. I disc lespedeza sod for wheat and drill on a hard seedbed so I have little invested if the wheat fails.

Cotton: New Madrid raises as much cotton per acre as any place in the United States. I like to save cottonseed out of my first picking to have it early. At this time the seed is often so green that I scatter it one foot deep over a big floor so that it will not heat and cause poor germination.

Lespedeza: It is my opinion that lespedeza is New Madrid County's best legume. Following are some of my reasons for thinking as I do. (1) This crop is not particular where it grows. (2) It is a heavy seeder. It re-seeds itself. (3) Its cheapness of seeding and other characteristics make it fit in rotation. In 1938, my best yield of wheat was twelve bushels to the acre. This same field yielded also one ton of lespedeza hay and in addition 750 pounds of lespedeza seed. This was three crops on the same land in one year, and the land could be seeded back to wheat in the fall for the next year.

<u>Alfalfa:</u> While I said previously that lespedeza is Southeast Missouri's best legume, alfalfa does well here. I almost never fail to get a stand of alfalfa if I put alfalfa in after wheat. I break the ground around the 4th of July and start disking in August. I try to plant between the 27th and 29th of August. I sow the seed on the disked ground and then roll it in with a corrugated roller. I roll it about twice a week until I get a rain on it to bring it up. For four years I have had ten acres of alfalfa and brome grass mixed. The first cutting was always about half brome and half alfalfa and made all the hay ten mules needed all year. All other cuttings were pure alfalfa.

<u>Sunflowers:</u> I raised sunflowers in 1934 and 1935 while the original AAA [Agriculture Adjustment Association] was in effect. Forty acres averaged 1000 pounds per acre at three cents. Sunflowers are easily raised, produce well, but may be worth nothing at harvest time. Then too, the harvest requires a lot of labor and comes when one is very busy. In case of heavy rains the crop may rot or fall to the ground.

<u>History of the Herd:</u> In 1928 my father came to the conclusion that beef cattle production would be profitable in Southeastern Missouri. He purchased a number of cows at the National Stock Yards, Illinois. These cows were largely typical old style Durhams. A few showed some modern shorthorn characteristics, and others showed Hereford ancestry. Some were black. These cows were turned in [to breed] with registered Hereford bulls from Jackson and Fredericktown. Incidentally, the polled bull from Fredericktown was the first Polled Hereford [a bull without horns] I had ever seen. This bull I traded off to Mr. Peck of Malden for a young bull which Mr. Peck thought was pure bred. This was a very bad mistake.

The story of this bull is one we will always remember. Peck took me out to his farm near Malden to see the bull in the spring of 1932, the year after I arrived here in Missouri. I wasn't particularly fond of his bull, which was only a yearling, but the bull I wanted to trade had been here several years, and I thought a change of blood would be good. Accordingly the trade was made, Peck to pay for trucking. Well, apparently the yearling bull Peck brought over was rather wild. When he came out of the truck, he was in high glee and never slowed down until he was lost in the woods two miles east of here. We hunted for him nearly a month before we found him and brought him back to the herd. He was always a high-headed, wild bull, and never worth a cent.

The original stock was heavy milking, old style Durham cattle and were purchased in St. Louis. One of these cows was a very heavy milker but had milk fever. When I desire to cull some cow, I will wait till she calves, then take her calf away and let some other raise the two calves while the cull goes to the fattening lot. I think heavy milking is very important in a herd cow.

<u>Advantage of Spring Calves:</u> Up to calving time the cattle use stalk fields. The exercise is very beneficial. When calves begin to come, the cows' pasture is wheat until about May 1 when they go on bluegrass until frost. I always have more lespedeza than needed for pasture alone since I thresh 100 acres for seed and cut some for hay. Of course all wheat has lespedeza in it. When I cull one, I can easily see which cow is not holding up well, and a shy breeder is easily detected before one wastes several months of feed. By having all calves in early spring, the calves are large enough to utilize grass well by the time the grass is ready. The advantages are: (1) The calves are wintered with less feed when the calves are dropped in spring. (2) The calves are nearly all castrated before they are a month old. When "fly" time arrives [the hatching of house flies and horse flies], no more calves are castrated.

Breed History: Polled Herefords are an off shoot from the horned Herefords. For this reason there are probably ten times as many registered horned Herefords as polled. It is likely that if one has $100 or $150 to invest in a bull, he will get a better beef individual for the same money by buying a horned individual. It has never been my policy to place polled characteristics above beef characteristics. In selecting individuals in my herd, if a horned individual is a better beef individual, she is retained. I believe if one is willing to pay $200 for a bull, one can come more nearly obtaining a polled individual. I am very cautious about handling and buying young bulls.

I have purchased three young bulls. All of them were fed with the milk cows the first year to make them as tame and gentle as possible and to make them accustomed to people. It is my opinion that a farmer should avoid purchasing any bull he cannot handle. No farmer should buy bulls that come from herds that show flightiness when strangers come around. Wild cattle are hard to handle. I believe, however, that wildness in cattle is as much the matter of handling as the breeding, but both factors enter in.

While on this subject of flightiness in cattle, I might compare the polled bull I traded to Peck and the Applegate bull which I kept until December 1935. The polled bull was easily driven where any one wanted but always could jump any fence to get back to the herd. The Applegate bull generally stayed where he was put. One could always bring him in from the herd at will. The polled bull dominated the herd from 1928-31, and I believe this bull put up some good jumping because my first year of experience with a herd in 1931-32 gave me the impression that we had the wildest herd of cattle I ever saw. Now this 1935 year after the Applegate bull has dominated the herd for four years 1932-35, we have a much quieter herd. It is my policy to ship all wild natured cattle because they make others wild. Also my advice would be to always be sure no break [in the fences] occurs when driving or cutting cattle. For this reason we bring the entire herd in to the mule lot whenever we cut out for any reason whatsoever.

The two bulls which I intend to use in 1936 and for the next few years are both double standard. Polled Herefords cost approximately $125 as year old calves. One came from the Goby herd at Marion, Illinois, and the other came from Holman herd of Salem, Missouri. The Goby bull has almost entirely a polled ancestry and should poll nearly 100% of his calves. However, he is a little more upstanding at present than the Salem bull. He has an exceptionally fine head and front quarters but is low in the tail head and not so good as I would like in the rear quarters. The Salem bull is very low set. He is out of a polled dam and polled sire but has two horned grandparents and for that reason probably will not poll as many calves. This bull seems to have a very fine gentle disposition. In my opinion he is a little more of a show bull than the Goby bull. However, I like the Goby full head best, and the Goby has the better color markings.

Keeping a prize bull was important before the days of artificial insemination.

The Hubbard herd of registered polled Herefords was started in 1936 with a top cow from the dispersion sale of the Turley herd at Knob Lick, Missouri. Two years later after the death of Lee Goley at Marion, Illinois, one of his best cows was purchased. No more females were purchased until 1935 when the reserve champion at the Cape Girardeau sale and show was purchased.

Some conclusions after seven years with beef cattle.
1.   One can produce top quality feeder calves to weaning age

at least as cheaply as one can buy them. If one allows $5 per acre for pasture, nothing for stalk fields and straw stacks, and an amount equal to pasture cost for hay, one would lose money producing any kind except top quality calves, because their cost would be the same and value less.

2. In 1938 the beef enterprise showed a return of $1.40 for each $1.00 of feed including pasture hay and corn. The profit came through in the feeding out so as to obtain top prices.

3. There are certain indirect returns from the beef herd which may make a farm profitable even though the beef herd is not given credit for it. Following are some of these: (A) Payments because of conserving crops for beef cattle. Except for the beef herd, I would not have received government benefits and made money on conserving acres. (B) Yields of corn and other grains following pastures are larger. (C) Manure returned also increases yields of crops.

Expenses must be kept down if a profit is to be made. I will endeavor to list some ways expenses may be kept low and explain some of my reasons.

1. Diversification in farming tends to lower expenses. It provides better labor distribution and thus teams are more fully utilized. Livestock in a diversified system makes pastures pay with low crop risk and expense.

2. Repairing tools should be done in a shop on the farm whenever possible. A small blacksmith shop may pay for itself each year on a farm.

3. In purchase and selection of tools, try to weigh carefully whether they will really pay. Many small farmers use tractors without really realizing how costly tractors are to run. Horses or mules might be far more profitable for them.

4. Don't run a tractor and leave a lot of work stock in the barn.

5. It is better to go to a bank and borrow than to buy equipment on time. One loses too much interest on time payments, and remember, there is always a discount for cash.

6.   An electric fence will lower fencing costs on temporary fences.
7.   Good farm business records will help hold down expenses.
8.   When repairs are needed, it may save to get them as soon as possible. A new plow point may save wearing out a grog. To lay planter shoes and plow points in time may save purchase of new ones. Remember the poem "All Because of a Horse Shoe Nail." It is very much the same way in maintaining machinery. Sometimes a loose bolt or absence of a needed bolt may result in breaking an expensive casting.

> MJS: For want of a nail, a shoe was lost.
> For want of a shoe, a horse was lost.
> For want of a horse, a rider was lost.
> For want of a rider, an army was lost.
> For want of an army, a battle was lost.
> For want of a battle, the kingdom was lost,
> All for the want of a horse shoe nail.
> —George Herbert 1651

9.   When one has teams which tend to run away, it is well to put them on the old tools whenever possible.

Chapter 3.
Farm History via Newspaper Clippings

Two articles appeared in the Missouri Ruralist about Ralph C. Hubbard's farming operation. Alma Hubbard's note on the article reads, "Ralph's being selected a master farmer was one real surprise. We don't know who was responsible for naming him. He received a very pretty "medal" on a watch fob. I'll have to admit I was pleased."

"Honor Five More Master Farmers,"
Missouri Ruralist, January 17, 1942

Ralph Hubbard, Lilbourn, is one of the youngest men ever selected to be a Master Farmer, being only 31 years old. Back in 1931 he was graduated from the University of Illinois and went

25

to his 945-acre farm in Southeast Missouri. This land was heavily mortgaged. By a program of diversified farming, including livestock, in a section where cotton had been the major crop, this farm became profitable for the first time. Mr. Hubbard has done such an outstanding piece of work in the management of this land that he could not be overlooked when final selections were made.

To young Ralph Hubbard goes a high honor and it is recognition to a new way of farming a relatively new country. Southeast Missouri was Missouri's last frontier, and Mr. Hubbard thinks it is a land of opportunity.

You'll go a long way before seeing better lespedeza than Mr. Hubbard grows, and he grows more that 500 acres of it. Here you see lespedeza on good land, and what it will do. In addition there is 70 acres of alfalfa. Other crops include 260 acres of corn, 135 acres of wheat, 80 of oats, and 62 acres of cotton.

The major rotation is a 5-year plan, with 2 years of corn, then small grain with lespedeza for pasture, wheat and lespedeza, then lespedeza for seed. A 7-year minor rotation is cotton for 3 years, small grain with lespedeza, and 4 years of alfalfa.

"Production of beef cattle and hogs have been my most profitable farm enterprise over a period of years," says Mr. Hubbard. "Wheat has been my least profitable, but lespedeza for seed has been twice as profitable as wheat, even on the same land. Cotton is my most profitable crop to the acre, but I raise only my allotment to comply with the AAA. Lespedeza seed production is even more profitable than corn."

His beef-cattle herd consist of 4 purebred Polled Hereford cows, and 77 grades. He gets a good purebred sire, one of the best. He has 43 purebred Duroc brood sows, and 40 commercial sows. All feed is marketed through this livestock. In addition he buys about 2,000 bushels every year, marketing beef calves at 15 to 18 months old in near prime condition.

This farm is well equipped with machinery, with 2 tractors, combine, hay press, corn-picker, threshing machine, corn elevator, and other implements.

Mr. and Mrs. Hubbard have an attractive farm home. It is

modern with many conveniences. They have 3 children, Martha Jean and Mary Ellen, who are 7, and Alma Ruth, 2.

Mr. Hubbard is active in community affairs, is trustee of his church, a Sunday school teacher, president of school board for 2 years, member of the board for 6 years, and a member of the executive committee of the Farm Bureau for 2 years. He also is active in several farm organizations.

"Swamp Cattle," Missouri Ruralist, October 26, 1940

Missouri's possibilities as a livestock state seem to be unlimited. The grasslands of North Missouri long have supported fine herds of cattle and sheep. Much of our crop land in all areas of the state have been turned to livestock production. Even our Ozarks have become cattle country following the introduction of lespedeza.

And now Southeast Missouri, the state's "Dixie Land," is finding that cattle, and hogs to a lesser extent, can fit with a program of diversified farming to good advantage. Southeast Missouri as a potential livestock country catches the imagination of any North Missourian whose pastures are often covered with snow and ice during the winter months. In Southeast Missouri year-around growing pasture is almost an actuality. Rye and barley, as well as the winter legumes, thrive throughout the open winters that usually prevail.

As a leading disciple of livestock in the "swamps" we give you Ralph Hubbard, of Lilbourn. Mr. Hubbard operates almost a thousand acres along one of the many drainage ditches that have been cut down through Pemiscot and New Madrid counties. His is rich cypress land that grows anything.

Twice since Mr. Hubbard has lived on the land—and he is a young fellow at that—he has been able to ride the range in a rowboat.

On this farm, Mr. Hubbard has a really fine herd of 80 Polled Hereford cows. Only a few of the females are registered, but virtually all are of purebred stock. By careful selection of promising heifers, he is building up a high-grade herd. When it comes to bulls, Mr. Hubbard buys the best registered animals, of course. He has raised several bulls from registered females that are top-notchers.

Mr. Hubbard feeds a good type of chunky, short-legged Duroc hog.

The sharecropper system so prevalent in that section is not followed on this farm. All labor is hired, although about 11 families live on the farm.

Following this system, the entire farm is operated as 1 unit and the fields are immense—running around 125 acres each.

A definite rotation is followed and, although the soil is originally fertile, Mr. Hubbard uses legumes and other good practices to keep it that way.

One of the prettiest sights on his farm this fall was a 130-acre field of lespedeza. We are most familiar with lespedeza as a poor-land crop. You ought to see it growing on this really fertile land. Here it certainly is not a hill crop with not a hill to be seen for miles and miles around. This lespedeza provides an abundance of pasture.

For hay there is a fine field of alfalfa and also plenty of good clover hay. Small grains, wheat, barley and rye provide good fall and winter pasture long after North Missouri is feeding hay and grain.

"This is a land of great opportunity," says Mr. Hubbard, and proved that he believes this thoroughly when he left the great farm section of central Illinois around Champaign to take up the operation of this farm. "We still have a great social problem," *he says, "but perhaps we can solve that some day." [The problem was that of segregation.]*

Mr. and Mrs. Hubbard are making a real contribution to their community. They have built a fine home on their land and take part in Community enterprises. Such folks as these on the fertile land of Southeast Missouri will make it one of the nation's greatest farming sections. Here we find livestock taking its place in such a program with cotton, corn, alfalfa, and the other many crops that will thrive here.

Mr. and Mrs. Ralph C. Hubbard
are landowner winners.

Plant To Prosper Contest, December 16, 1938

The Plant to Prosper Competition was a contest sponsored by the "Commercial Appeal" and the Memphis Chamber of Commerce Agricultural Committee with the cooperation of the Agricultural Extension Services of Arkansas, Mississippi, Missouri and Tennessee. Alma Hubbard entered the competition by filling out a detailed inventory of buildings, equipment, livestock, feed, seed, and food products that the farm produced, plus a personal narrative. Her written narrative puts wings on her dreams of possibilities for the farm and documents what the farm has produced.

Personal Narrative by Alma Hubbard

Your Life On The Farm: Ralph Hubbard, twenty-one and just graduated from the University of Illinois School of Agriculture, came to this farm in 1931. Two years later he married Alma Heath. I, Alma, had lived in this community since early childhood. My college work was done at the State Teacher's College in Cape Girardeau, Missouri. I taught three years in a rural school. In 1934 twin daughters, Mary and Martha, were born. Ralph is a farmer by interest as well as by education and circumstances. The home is on the farm by choice as well as necessity. We started farming during the "dark days" of the depression. At first it seemed we would lose part of the land, but we had no money to lose and perhaps a farm to gain. Now we are happy over our progress. Our dream is to have a pure-bred stock farm—Polled Hereford cattle and Duroc-Jersey hogs. We have a grade herd of both that has topped the market several times this year and are starting a small herd of choice purebreds. We have always complied with the government program, thus hoping to build our soil as well as pay for it.

What the Plant to Prosper Competition Has Meant to Me and My Family: It has been a challenge to plan and direct our ability toward better farm living and the resulting better farm paying. It renewed our zeal to lift the debt from our land. It is not only a challenge but also a stimulant to urge us to give our home and farm the best management we could. Although we have kept a Farm Account Book and a Home Account Book for seven years, this contest caused us to be very exacting, and all estimates are taken from actual records. It has meant we farm wives can get more co-operation in our home improvement work--who said we couldn't be interested in such?

Narrative of How I Lived At Home in 1938
The worry of a mortgage is depressing. This fact and this contest caused us to plan carefully so as to utilize all possible (if

profitable) chances to lower our cost of living (without lowering our standard of living). The following will illustrate how the cost of food was lowered by utilizing the farm's supply.

1.   Flour: exchanged wheat for flour at a lower than retail price.

2.   Meat: (a) chickens, used at least one per week, (b) pork, butchered three hogs, cured part, canned 75 quarts, lard for our own use, sold $3 (of lard). (c) beef, fed our cattle but do not find it profitable or satisfactory to kill for our beef supply.

3.   Honey: all we cared for, used some for gifts, exchanged $25 for other groceries. By exchanging produce for groceries, we get a better price. The honey is wrapped in a special cellophane wrapper.

4.   Fruit: (a) fresh from our orchard. We had apples, grapes, strawberries, cherries and plums. Sold $6 of grapes. (b) canned enough for 250 quarts, used money from grapes to buy peaches and pears.

5.   Milk: pure-bred Jersey cows furnish all milk, butter, and cheese we could use and sold $20.

6.   Eggs: all we needed and exchanged $60 for stable groceries.

7.   Garden: An excellent spring garden. We had five vegetables at Easter. Nice supply of asparagus. Canned 200 quarts of vegetables from our garden. (I have no household help so my mother kept the twins at busy times, and in exchange I furnished her all the vegetables she needed; thus the garden was for six persons.) I planted three complete fall gardens, and all failed to "prosper" due to adverse weather. At present writing we have tomatoes, peppers, sweet potatoes, turnips, radishes, and mustard greens. When I didn't have vegetables or fruit from the farm, I bought them. I didn't sacrifice balanced meals for my family's health is of major importance.

Chapter 4.
Exploring Friendships, Alma's Letters (1936-1949)

For some thirty-six years, my mother, Alma E. Hubbard, wrote in a Round Robin, sharing her life with nine farm women in various parts of the country from Wisconsin to Texas. They assumed the names that they had used in a letter to the "Household Circle" in the *St. Louis Daily Live Stock Reporter* —Alice Ben Bolt, Mandy, Hibiscus, Molly O'Neill, Swamper, Missouri Shepherdess, Happy Rancher, Ex-Steno, Jersey Fawn, and Blossom. You can guess that Alma was the "Swamper" in Southeast Missouri. Someone in the group contacted the other writers and invited them to share their farm interests with each other. The letters give a running commentary of life on a farm. It started before the days of electricity in Southeast Missouri and ended when the twins were thirty-eight years old.

Alma had some notes in one of her "Memory Books" about the "Ladies of the Robin." The list includes:

Alice Ben Bolt (Mrs. H. S. Rule, Linden, Wisconsin)—
    lives on a ranch, "garden of God's country," a
    writer.

Mandy (Mrs. Myron A. Hickok, New Windsor, Illinois)

Hibiscus (Mrs. Ira B. Eisenbise, Lanark, Illinois) Edna,
    older, raised lots of flowers, cans lots.

Molly O'Neill (Mrs. Verlin Snouffer, Paola, Kansas)—
    Interested in 4-H club work, living on large rented
    farm, all children grown, home burned, planning
    to build.

Missouri Shepherdess (Elora Blaetner, Wyaconda, Missouri)
Happy Rancher (Mrs. J. R. Jeske, Alamo, Texas)—had cattle ranch, Bingham, Nebraska, moved to Texas for Dale's health, has fruit farm, children—Glen Royal, Paul Allen, Dale Lloyd, Mary Joyce, Emma Jayne.
Ex-Steno (Mrs. Harry Ostergard, Gothenburg, Nebraska)—children: Larry Dean, Dick, Jim, Jack, live on ranch.
Jersey Fawn (Mrs. J. Bintzler, Naperville, Illinois)—near Chicago, likes to knit, baby 13 months.
Blossom (Mrs. Hans Bjoedal, Centerville, South Dakota)
Swamper (Mrs. Ralph Hubbard, Lilbourn, Missouri)

The first two of Alma's letters are presented in their entirety. Subsequent letters have been excerpted. Comments by the author are noted by MJS for Martha Jean Hubbard Stewart.

Writing in a Round Robin appealed to Alma's literary interests.

Dear Robin,   *November 1, 1936*

How very nice of you to fly my way and bring me the interesting letters and pictures. Thanks to you "Ex-Steno" for that. This is my first robin in the "Household Circle." How excited I was when it arrived. Reading the letters seemed to give me renewed energy. I was just feeling tired and stiff from having "overworked" the previous day—wash day. I usually over do the thing on wash day. (Ralph says I try to do everything on wash day.) Are you like that? Some days I seem so full of pep, I just want to do all the things that are waiting; and the more I do, the more I feel I must do before the day is over. The next day I'm tired and "pep-less." It was thus the robin found me, and I felt like "flying" about again after I read his message. With twin baby girls I find plenty to do every day.

I'm hoping all the wee ones have arrived safely. Don't we all love a baby? My family consists of twin daughters, age two year last February 12th. What a joy they are to us, and how they do enjoy each other! Of course they talk now--and how merrily they chatter all day! They are learning Mother Goose jingles now—know about twelve—these they try to sing. Maybe you would like a word picture of them. Martha Jean is—and has always been—larger than Mary Ellen. They are not identical, but both are very blond (like their dad, I am brunette). Jean is athletic and graceful, good imagination and expressive at home, but very shy with strangers. She is—and has always been—more mature than Mary. She takes care of Mary and "bosses" the play. Mary is prettier because she is fatter, more chubby. She is awkward, happy-go-lucky, good-natured baby, always at ease and laughs and talks to everyone. They are just two entirely different characters—as different as any two other sisters.

Twin girls are a handful of squirm.

At first I was sorry the girls were not identical—they are so cute when they are alike. Now I am glad they are different. They are more interesting as they are, to watch the contrast in their mannerisms. Then, I believe they are more likely to be normal and individual. They do not have the same faults or same virtues; and thus, they will more nearly balance the influence they have on each other's character.

Some of you are commenting now, "Twins aren't much more trouble than one," others, "Twins are double trouble." I say they are likely to be double trouble, and sure to be more than double fun! They are likely to be more trouble, for, as in our babies, they are often frail or premature. My girls slept in wool blankets with hot water bottles at their feet and side, a fire in the room night and day, for six weeks. They were fed at two hour intervals, and each had a different milk formula (which must be prepared just so). They kept some one—sometimes three of us—busy night and day. We look back and smile at the rate we sailed around those first months. My mother lives just next door, fortunately for me. She has been so much help—even now she usually dresses them while I dress, as we usually go together; I drive, she doesn't. Mother and Dad always come to see the girls every day. How they enjoy them, for I am an only child.

Wait! Wait! Interruption! (Continued on November 3)
****
Election Day! Gee, what a gloomy, dark, muddy day! And we live on a short stretch of "gumbo" road, so it was bad getting out to vote. Ralph was a clerk, so he went early. I'm not radical in my politics. I only hope the right men are elected today. Certainly lots of politics on the radio now. I'll welcome a change in programs.

MJS: I remember my father saying that he had refused to sign the poll books because a deputy sheriff intimidated voters by parading up and down with a gun in front of the election polling place at the school. My father was from the North, and he was not acquainted with the southern way of discouraging African Americans from voting. This was before voter registration.

My mums are pretty now. I hate to see them go. I had yellow mums in a blue pottery vase for my centerpiece last evening. I had a party of eight women at my home for a "Wear-Ever" aluminum demonstration dinner. Perhaps most of you are familiar with this. The demonstrator furnishes food, cooks, and serves dinner, then washes dishes etc. all free in the demonstration. A delicious dinner without our help surely seemed a treat to us farm women! They cooked carrots, beans, cabbage, rai-

sins, apples, and potatoes without water and minimum loss; and roast and steak and pancakes without lard. Ralph really beams when pancakes are mentioned.

Our climate permits us to raise several crops so we have had a hit this year on some. We raise wheat, corn, cotton, oats, hay. . . [The last page was missing.]

[A side note in the margin] Molly O'Neill: our local dime store had such a dish as you mentioned in pink glass on special sale 10 (cents). Perhaps they have yet. Shall I see? Your baby is "mighty sweet." Thank you so much for offering to see about the dish. Unless I can't get the clear old fashion crystal, I'd rather not have colored.

Alma and her parents take the twins on an outing.

January 10, 1938

Happy New Year to all!

The girls are tucked in bed, and Ralph is listening to the Illinois-Purdue basketball game; so now I'll try to send the robin on its way.

It was nice to have the letters and especially the good pictures. It seems, as my letter was lost, some would like to be introduced again. Perhaps this will help. I am three years from thirty (27), rather tall and slightly plump, have dark brown eyes and hair, and fair with freckles in summer. I have been married five years, and my family consists of twin daughters age four this February 12th. They are very blond—like their Dad. I am an only child of a farmer, after college taught three years, and married a farmer. My husband had prepared as an agri. teacher at Ill. U, but instead of teaching, came to Mo. and took a much-mortgaged farm. This farm joins my father's [Joseph M. Heath], so I live across the road from my parents—I appreciate this.

 Now about our part of our country. We are in the extreme southeast part of the state, near the Mississippi River—twelve miles east of us. There are thousands of A. [acres] of cotton in this corner of the state. In fact the highest yielding cotton per acre in U.S. is in this area. When prices are good, this crop brings lots of money in our country. The crop requires so much hand labor that there are many homes per sq. mile. The land is mostly in hands of large ginning companies or plantation systems. You see many poor tenant homes, then a mansion of a plantation house. This is especially true when you drive from here to Memphis through Ark. Cotton is developing this country, but it will eventually ruin it; as they seem to just grow cotton, cotton, until the soil is ruined. We have had a great tide of southern planters and Negroes the past three years. I don't like cotton. It requires so much hard labor, keeps children out of school, and leaves such ugly fields--I am glad to have alfalfa near the house. Does my picture seem dark? It isn't all like that and many like cotton. Our farm is 945 A. and only 40 A. in cotton. We raise corn, wheat, hay, and lespedeza for seed. We have 150 Herefords and raise from 300 to 500 hogs per year. Ralph likes livestock and perhaps we shall sell some day for a cotton plantation (when the price gets right!). We have a consolidated school system (2000 enumerated in the system) and transportation to a central school. (Altho Dad and Ralph are on the "Board," the transportation system doesn't suit me. I don't want my girls to ride 10 miles to and from school on an over-crowded bus over not-to-good roads.) Now am I completely introduced?

 Hibiscus, I looked you up in the Atlas and found you're quite a distance away. My "second love" is flowers too, but I'm not as successful as you. To work with flowers always chases the "blues," doesn't it? Cypress trees are native here, as this was a sunken swamp. I have a pretty one in my front yard. There are also many wild pecan trees. Would you like to exchange quilt scraps for seeds?

Ex-Steno, thanks again for including me. I knew the picture was you before reading the back. You look like your letters—vivid or rather vivacious, full of "pep" and direct. (Please don't object to my stumbling adjectives, but I do like the picture.) The boys are fine too. Do come often to the "Household." I watch for you as you are my "god-mother." We are near New Madrid and twenty miles from Betty Glad.

Happy Rancher, yours is a sweet family too. I enjoyed seeing their bright faces. Your Dale looks like his mother. Doesn't he? I had a friend at Amarillo, Texas, and liked her description of the country. I hope the fruit farm is a success.

Alice Ben Bolt, I enjoyed the description of your country. The lost letters don't matter—we are all here again.

Enough for me this time! Ralph has finished with his ball game, and now I'd better help him pack. He is driving to Urbana, Illinois tomorrow—340 miles. He is going to visit his mother and brothers, and also attend "Farmers' Week" at the University. Will you be there "Jersey Fawn" or "Mandy"? I find "Farmer's Weekly" very helpful, but I'll not be there this year.

Please fly my way again. Perhaps I'll have a picture by that time.

Love to all, Swamper

April 24, 1938
Dear Robin,

This letter will probably have many errors, as I am trying to get the "typing habit." My Mother-in-law always comes to visit us for the Easter vacation, and instead of giving us a Christmas gift, she brings a much desired typewriter. We were wanting one but didn't feel like getting it just now. She must have read our thoughts.

Easter really found spring in southern Missouri this year. I *had asparagus, radishes, lettuce, spring onions, mustard greens, spinach and a few strawberries in my garden. There were roses, snowballs, weigela, honeysuckle, spirea, and tulips in bloom in the yard. My garden peas are blooming and tomato plants are four inches tall. Most of the fields are plowed. We probably will begin to plant corn this week.*

We are twelve miles west of New Madrid, our county seat, one of the oldest towns of the state, and of historical importance as it was settled by the early Spanish explorers. This corner of the state sank in time of the earthquake in 1811 and was left as a marshy swamp until the idea of drainage was tried, about thirty years ago. Now it is a level valley drained by parallel dredged ditches a mile apart. Our big worry is spring floods—it makes me think of Holland. We have a few cypress trees left to remind us of the days of the swamps. I am rather proud of the one standing on my front lawn—even if it isn't in the logical place. The fern-like foliage is pretty in summer, and it is always covered with loads of seed cones or balls.

One place of interest near us (I like to visit) is an old pottery. Handmade pottery is made here on a crude, old potter's wheel by the family that has been working at this trade for six generations or more, at the same location. They make lovely candlesticks, lamp bases, flower bowls, etc. of mottled blues, yellow, and brown clays. They also had an exhibit at the Chicago World's Fair in the Missouri Hall.

MJS: This pottery is located at Bloomfield, Missouri, on a slick of clay from which the articles are made. When we were children, we would stop with our parents and traipse through the rustic buildings and yard that housed the pottery. Occasionally we would catch a potter at work with his wheel. A lump of mottled clay would be coaxed by the potter's hand dipped in water. His feet would pump the wheel that set the clay to spinning in a dizzy whirl. Once we visited the kiln, a brick building built on the premises and fired by logs. We learned to practice patience as we waded through the weeds to find the right hidden treasure. Today as I walk through the mall and see the displays of pottery,

I think of the pottery embedded in the grass and weeds and the primitive way an artist worked an old clay slick in the Ozark Hills.

. . . Molly, I enjoyed hearing from you. . .Yes, I think Edward did wrong in marrying a divorcee, but perhaps his loss was England's gain! If his emotions are so unreliable, perhaps his influence would be too.

MJS: Mother commented on the news of the day. The world was atwitter with Prince Edward giving up the throne of England to marry an American divorcee.

I—rather by accident—took a course in "shop work" in college, and since that time have rather enjoyed refinishing and repainting furniture. With varnish and brush I have been trying to repair the damage the twins did this past winter to my furniture. It is time to clean windows, but with garden and chickens I haven't had the energy to tackle them. Here is something we did as a money saver—maybe you would like to try it. With three small pulleys, a cord (clothesline), and some awning material from the mail order house, we made a ten-foot shade for our front porch that cost about $2.38. It really works and is pretty.

MJS: "The mail order house" was the catalogue for Sears & Roebuck. Every family looked forward to the big catalogue that came in the mail for the season. Catalogues were thumbed through repeatedly for ideas and items. It was a reference book that documented new products. Money was so scarce that it was often called the "wish book." Decisions about purchases were pondered for weeks.

Mandy, our bathtub system is much like yours, except in a small room just off the kitchen. We don't have electricity and so do with substitutes. The one I could not do without is my Electrolux (oil-burning refrigerator). The initial cost is high, but they are just fine in every way. I think hard-working farm families deserve a few nice things to enjoy in their homes. This talk of

41

jokes on the baby question brought to mind a joke that occurred in our family this winter. The younger generation consists of girls in the Hubbard family. Naturally everyone was eager for a boy. A neighbor gave the "twins" a Persian kitten. Ralph decided it would be fun to write his favorite brother a letter telling all about the kitten, but calling it an "addition" and making it sound as if we had a baby boy, until the end of the letter and then reveal he was speaking of the kitten. He had loads of fun thinking of the envy and excitement. But—he failed to make it clear enough at the close, or Bob was too excited, or too English to catch the joke. At any rate, Bob took it seriously and spread the glad news to both his and Ralph's embarrassment. Ralph's Mother even sent some hand crocheted booties! No more such jokes in this family!

Hisbiscus, . . . I hear my family arriving. The girls had gone with their Daddy to see if the alfalfa is ready for the first cutting. The girls have grown so tall this winter that they are no longer babies. They have loads of energy and a wonderful imagination. There is a constant line of questions—they ask, "why for" concerning every topic. Their newest achievement is learning to "fistle" (whistle). Now they are trying to learn to wink "one eye." Wouldn't this world be a dull old place without children? . . .

P.S. Aren't we proud of Alice Rule and her "Rural Chats"? I've always had a secret longing to take journalism, but I guess all mortals long to express themselves in writing at times . . .

November 23, 1938

This is a lovely, warm, autumn Sunday afternoon. This autumn has been a rare and delightful season. This is Thanksgiving week, and the petunias are blooming gaily in my window box. (Isn't a window box a paying bit of garden—being out of the reach of children, chickens, and dogs?) I don't care about house flowers as my living room is a north room, but I do have two pretty ivy plants in my kitchen windows. They are the easiest kept plants I have ever tried. I enjoy the flowers in the yard, but we haven't had much money to spend on such, so my flowers are mostly cuttings given by friends. We did buy two cedar dedorus evergreens, and I was amazed when one of them grew all of fourteen inches this summer.

Mandy, you asked about the pottery prices. I'd call them reasonable. Some call them high, but they are certainly more reasonable than the same thing would be in a department store in the city. For example I paid $1.75 for a matched set consisting of candlesticks and candy dish with a cover. They are a fawn, with mottled lines of blue, red and brown. I don't tire of using them as I can use any color scheme by changing the color of the candles. Large heavy bookends were $1.00; vases and flower bowls were 50 cents to $1.25 depending on the size. The best work this potter does is that done on the potter's wheel as lamp bases, flower bowls, vases, and candlesticks. The bookends and plaques are rather crude.

The twins had company today—a little girl from Sunday school. I am afraid they almost played her down. She was so tired when she started home—and she is much older than they. I have always worried because they are slender and not very plump. Today I decided they just run all their fat off. They never take a nap now, but they sleep twelve hours at night—six till six. I don't know just what I'll do with so much energy when winter comes and they are forced to play in the house on bad days. Martha knocked her tooth out on a slide, and it gives her an amusing lisp when she is trying to talk at her usual high speed, and a quaint air when she grins. I try not to notice as she is sensitive about it.

Oh, I must not forget to tell you a bit of luck this summer. I won a $50 prize for naming a farm program coming over KMOX, St. Louis—no box tops to send either. Was I bowled over? No, I am not a radio contest fan and haven't become one since winning. (Ralph said I'd spent it on stamps entering other contests—he was wrong.) We listened to the program and talked of the naming contest, and I entered on sort of a bet. They sent a special delivery letter and telegram and offered to pay our way if we would come up for the radio presentation. We couldn't and didn't care to. The check was mine, and I felt real liberal—it was like a gift out of the blue---so I just broke out and bought a Master-made Sellers' kitchen cabinet and new linoleum for the kitchen and bathroom. The cabinet was slightly damaged when it arrived, so they knocked $5 off the original price. I took this and bought a small, modern-style bookcase for the living room. Did I stretch that $50 and get a lot of pleasure from it!

There are eighteen Women's Extension Clubs in our county, and one of our county projects this year was kitchen improvement. Each club chose a Kitchen Improvement Demonstrator—it fell to me in our club. There is to be a county tour of all the kitchens of demonstrators this month. My new cabinet is white with a black base. I repainted my kitchen woodwork white, and built-in cabinets have a black base too. I like this idea of a black base, for it doesn't show the toe marks. My walls are ivory, and the linoleum is ivory, black, and green. Mother says the kitchen is the brightest room in the house; it should be, I stay there most of the time.

"The Commercial Appeal" sponsors a contest, "Plant to Prosper," among southern farmers to stimulate better methods of farming and home improvements and living. Farmers in southern Missouri, Arkansas, and Tennessee are eligible. You turn in a record book of all expenditures, foods, supplies by the farm, history of the farm, soil conservation program, etc. We won in our county landowner contest, and thus are invited to a banquet and tour of the city of Memphis, Tennessee, on December 15. Are you calling us "contest bugs"? It was fun and beneficial as it made us watch accounts.

Just now we are hoping—with our fingers crossed—that we may be on a R.E.A. (electric line). The first thing I want is a toaster

As we near the day of Thanksgiving and read of the turmoil in Europe, I am indeed thankful that at least an ocean separates us from the present situation.

MJS: The turmoil in Europe led the Americans into World War II. In the isolation of the farming communities our lives center on the farm. Farm work required much brawn and was all consuming. Only the radio and the weekly paper kept us in touch with the outside world.

Just as I was finishing a large washing, a judging team composed of three men from the "Commercial Appeal" office, a woman and two men from Missouri University, and our County Agent arrived and announced we had been selected as one of the three highest in the Missouri Contest "Plant to Prosper." Was I flabbergast! I hadn't thought of having a chance in the state. We won't know the outcome of the state judging until Monday of next week. They just insisted in taking a lot of pictures of us. I almost wished I'd not entered the contest when I think of those pictures. That rattled me until I couldn't concentrate enough to finish this.

Hoping the coming holiday seasons bring <u>you all</u> much cheer. (Are you smiling at the "you all"? That is used a lot in Missouri, and my college English teacher insisted it is correct. It just came out that time before I thought, so I'll leave it.)

August 21, 1939

Oh, me, I've had the robin three weeks, and each day I've planned to write. Sorry. I've just canned and canned-- grapes, tomatoes, peaches and apples. Now, while my family is sleeping, I'll write. There is a little ivory bed in my bedroom, and there sweetly sleeping on her tummy is a little redhaired girl (not really red, but a golden brown). She came to our house June 16th, and now at two months weighs 12 1/4 #. The twins are so excited and delighted with her, and her Ma thinks she is a dandy lady! Her name is Alma Ruth, and she looks very much like my baby picture. (Poor child!) We surely enjoy the baby. The twins were so much work and worry; we didn't have time to enjoy them their first year. Ruth should have been our boy, but we are happy to have three little girls. (The twins wanted a girl, so they were satisfied.) All of the last generation in the Hubbard family was boys (five of them), and this generation is girls. It seems the name is going to end at the present rate! Ha.

Two Almas pose for pictures – Alma Ruth and Alma Etta.

MJS: The night Ruth was born, Mary and I were shipped off to Memo and Granddad's house. (Mattie and Joseph Heath) It was not unusual to spend the night with them. The next morning while Memo was cooking breakfast and we were clattering, she asked, "How would you like to have a little sister?" After a long pause while she wiped a pan, she added under her breath, "Well, you have one." We didn't believe her, but I was ready to go home and check it out. I remember bounding into the house. Mother was in bed, and in the corner of the room was a basket with a baby, swathed in a white gown and white blankets. Her little fists were doubled up and tucked under her chin. She was fast asleep. Momentous occasions heighten perceptions, and I can recall today the sequence of events as if it were yesterday. We stood and looked. It was more than I could take in. Then we went outside to swing on our rope swings in the back yard. I remember telling Mary, "Our whole life has been changed. We'll be different now. Baby Ruth will be a part of everything we do."

This has been a pleasant and cool summer here. We have had plenty of rain and crops and gardens are good. Ralph did the gardening this year and has enough tomatoes for six fami-

lies. We have been enjoying tomato butter. I make it by adding a sliced lemon to four cups of thick tomato pulp (seeds removed), and two cups sugar, and cook for half an hour. We like it better than the preserves.

You asked about the contest (farming and home making). Yes, we won a $100 check as winner in the landowner division in our state. We attended the luncheon banquet in Memphis last December. Over 600 were at the banquet; 27,000 were in the contest. This contest is sponsored by the "Commercial Appeal" and is directed by our county agent. No, I haven't entered any other contests, as my living has been rather quiet and uneventful awaiting the arrival of Ruth. I wasn't allowed to ride in the car for four months, and added to that I had a fungus growth on my finger. This had to be cut off. Ah, it is nice to be myself again and be able to work and go about some! Have gone to Sunday school three Sundays now, and Ruth has slept quietly in her carriage while I taught my class.

MJS: Doctors made house calls in those days. Our family physician came prepared to cut the fungus off Mother's finger. Mary and I stood close by to watch. All went according to plan. The doctor mashed up some pills of dry powder and then started cutting. It was a bit bloody, however, and Mary added excitement to the occasion by adding, "I feel like I'm going to lay me down." She was promptly removed from the scene before she could faint.

A few bolls of cotton are opening, and soon schools will be closed. Men, women, children, old and young will work in the cotton. It is pitiful to see it ruin children's chance for education. A cotton vacation is just a remedy, not a cure for the evil of cotton picking! How great the cost for the little it returns. The coming of cotton here has brought many southern plantation people. The omission of their "R's" makes their conversation amusing and interesting—their voices soft, rather musical.

Hibiscus, would you be interested in growing a cypress tree? I could send you some cones or "balls" from our trees.

47

There are just loads of seed cones on the trees, and they germinate easily here.

Yes, Hereford cattle are Ralph's favorite topic. I've even read a few copies of "The Hereford Journal," but not recently for I can't even get time to read my "Good Housekeeping"!

Lucky Hubbards!
Twin calves were born.

Our calves were nice this year—76 calves from 81 cows. But what of the price this next year? Recently Ralph topped the market on a load of fat steers. I'm hoping that means a new living room rug for me—goodness I need it.

I couldn't can earlier in the summer, but have made up for lost time recently. All jars are full except twelve quart ones. I canned grape juice in jugs—gallon vinegar jugs with mason tops. I fill the jar half full of grapes, add sugar and water, seal and cook in a water bath until the water boils. This is not ready for use for about three months but is delicious and makes such a pretty color of red. I cook them in a laundry tub in the back yard, so it wasn't much work and made my grapes go so far. I have 120 quarts of grape juice.

February 19, 1940

I was beginning to wonder what had happened to you. You were here in February, and here it is the first day of December. Do your shopping early—and I have. We just took the Sears and Montgomery Ward catalogue[s] and made out an order and hope that this year will find us serene and calm at Christmas. Maybe?

If this is full of errors--don't wonder—it is because I am suffering from a powerful headache due to intemperance! Sh! Don't shake your head that way. I didn't mean anything with *alcoholic content. I mean I am the sort that never wants to do a thing half way—when I can, or sew, wash, or read I usually get so interested I overdo the thing. This time it was "cut-out" embroidery. I did it every spare minute and then some, and now I am having a return of the old eyestrain I suffered when teaching. I bought linen and a McCall's transfer pattern and am making a set of scarves just to fit the dresser and chest in my bedroom. (Since this is Ruth's room too, and naturally the family congregates there to play with baby, the furniture gets plenty of bangs.) This furniture was a fifth wedding anniversary present, and I don't enjoy seeing it scratched. I do think the scarf is pretty—a poppy design done in buttonhole stitch—all white, with portions cut out to form the design. Have also made some new comforters and the twins four new dresses for spring. We have had an elderly woman to help with the housework this year, so I have a little more time than I ever had before. She isn't very effective, or expensive, and I do like to have a little time to play and read with the twins, for next winter they will be gone to school. They celebrated their sixth birthday February 12. Many send their five year olds to school. I am not anxious for my girls to grow up. I have taken a little time and taught them at home, and they are about as far along as the ones that have attended school, but I am rather puzzled about starting them on the piano. I play simple folk songs and hymns, but I taught myself after I had a few music courses in college, so I don't know just how to proceed from a child's viewpoint. Ralph is more help than I. He had lessons as a child and then played clarinet while in University.*

We, too, have had an unusually hard winter for this locality, and it has brought more than the usual share of flu and pneumonia. Our houses—most of them minus basements and central furnace heat—are not built for below zero temperatures. We burn wood in a range and a heater and try to heat a part of a seven room house. It was a small part when it was six degrees below. As the result of the sudden drop in temperature and the flu, our Ruth had pneumonia. We spent a tense week, but she recovered quickly. The fever was gone the third day. The doctor's treatment was very effective, but I was surprised as he did not want any salve used on her chest. I have always used lots of Vicks for colds. She had only one medicine, a white powder. Our neighbors had a little girl just Ruth's age, and they lost her with pneumonia. We read so much about pneumonia and its high death rate that I was surely frightened when the doctor pronounced it pneumonia. The death of our little neighbor reminded us that we have been fortunate, and God has been good to spare our baby. As a result Ruth has received more than her share of attention and is quite a spoilt lady now! Their doctor gave shots or serum.

MJS: Mother stood at the window, held Ruth, and watched as the family from a farmhouse in the middle of the farm walked through our yard. The women were carrying handkerchiefs and crying. Men carried the small casket. I stood at the window with Mother and watched. I didn't know why Mother was crying. She explained that we would cry if Ruth had died. Death was so much a part of life on the farm. Baby chickens died, baby pigs, baby calves. That was the first time I realized the fragility of life.

Alice Rule, I even knew a family so afraid of the word "bull" that they said "gentleman cow." As for the owl trouble, ours has to be 'possums. We have caught three in the chicken house. They crawl under an old bin. Ralph grabs their tails and drags them out. One of them we ate, and it was good too. They have such pretty fur, but I understand they are not worth much. I often wish I knew how to make something from them.

Mandy, I'd enjoy a peep at the new furniture. I have moved out my old dining room furniture and with the desk, the twin's table and chairs, the radio, plenty of rockers and have made another living room out of the dining room. We live in this room in the winter as it is the one with the heater in it. I have a new, walnut, gate-leg extension table in the living room for company dinners and the old stand-by out in the kitchen. I'll not get the chairs until my gals grow a little. I enjoy having the space the old table and buffet occupied. I hate useless pieces of furniture, vases, and whatnots. It's the plain modern furniture for me, for I believe it gives more space. I love antiques, but they are more in place in large, high-ceiling rooms.

Molly, I know the butchering means a lot of work, but is it good! Just think, we raise so many hogs and did not butcher a one. We use very little lard. Ralph had a nasal operation in December and wasn't so well, so we let it go. Does he hate to butcher, and I must admit we are very poor at curing it. Do you have an especially good method?

Alice, isn't it nice to have the family play games? Ralph is from a family of five boys, and they play just everything—checkers, ping-pong, horseshoe, chess, anything. I am an only child and must admit games are not as fascinating as a good story to me. Ralph and the twins are always playing something. He is really anxious to teach them chess.

The current was turned on our R.E.A. line on Christmas Eve. A nice present! I appreciate the iron most, Ralph the radio, and the girls are most curious about the Electrolux sweeper. It was my Christmas gift, and with all its attachments it surely swoops up the dirt. We are going to go ahead with our Electrolux kerosene refrigerator. Our minimum rate is $2.50. (a monthly rate for electricity)

MJS: I knew Mother was excited, but not until we went through the house and switched the lights on and off, did I understand what a difference electricity would make in our lives. The electric lights were so bright, for we were used to the Aladdin lamps with wicks that were constantly replaced. How easy it was to read anywhere in a room! We had always congregated around the table near the Aladdin lamp. At age six, I could not leave the light switches alone.

I'd better stop and see about soup for supper. We have soup about five nights a week in winter—milk soups such as celery, tomato, potato, and oyster.

December 1, 1940

This was just a usual Sunday with us. We drove to a small town (Lilbourn) about ten miles away for church and Sunday school. Ruth accompanied us—she usually stays with Mother and Dad (Mattie and Joe Heath). I teach the pre-school children, so she was no trouble. She was sleepy when the sermon began, so I took her to the car for a nap, and I read. The church is a weather-beaten, gray, wood building, and we are trying to get enough funds for a new building. This is the first time we have had a full-time pastor, and he is a great asset. After church we drove home to the usual chicken dinner.

Do you Gals belong to Extension Clubs? I do and enjoy the associations and learn so many things. We study and have demonstrations on such subjects as hooked rugs, milk dishes, salads, yard improvement etc. Of course our leader is the Home Demonstration Agent. This year my club sponsored a W.P.A. [Works Progress Association] hot lunch kitchen for the school for about a hundred pupils, also a W.P.A. gardening project. We furnished the land, seed, and cans. They furnish the labor to cook and serve the lunches and enough surplus commodities to give the dessert and bread. Yes, my opinion of W.P.A. labor is about like yours and all other red-blooded Americans' reaction, but if money must be tossed out to people for half a day's work, I thought our school children, many coming on a bus, might just as well have the benefit.

I have just finished making Ruth a blue wool Shetland cloth set—bonnet,

Ruth wears her scalloped cape to pet Alma's cat.

coat with little cape, and leggings. It looks nice, if I am bragging. I made velveteen leggings to match the twins' new coats. The coats were pretty, but no leggings. I was able to match the velveteen of the collar exactly. Our doctor says T.B. is on the increase among children because they go so naked. I don't see how they survive going naked from the hips down. Maybe this is a pet peeve of mine, but I say if God had intended for us to go naked, he would surely have given us fur or feathers. My sister-in-law, living at Urbana, tells me that a college professor's little daughter came to school in a lovely white fur coat and hand knit mittens, with her legs bare during zero weather, that she suffered from frost bite. I was surprised to find she envied me, getting to send my children to a rural school!

The Home Demonstration Agent made over 1800 mattresses for low income farm families in this county last year. Of course the idea was to use surplus cotton. These handmade mattresses are lovely—made of the finest cotton, not the rough kind of cotton you usually get when you buy a mattress. They are worth about $15 and soft somewhat like a feather mattress. At a little mattress shop where mattresses are made, I bought innerspring ones for the twins' bed for $9.50 each. This seems very reasonable to me. I know how to make a mattress as I have helped in making one in demonstration, but believe me it's some job! After the cotton is in the tick, you take a broomstick and beat it, and beat it some more.

MJS: The Extension Clubs, organized by the land-grant colleges and universities, were a boon to the life of the farming community. All farm families were instructed by the County Agents and Home Demonstration Agents who touched the lives of the farming community with their expertise and the latest in farm research.

Ex-Steno, what wonderful news! Twins are oh, so much work (as you know by now), but so interesting. My girls weighed six and seven pounds at birth, but now the six pound one weighs ten pounds more than her sister. One is serious, the other a tease. They are very loyal and always insist on going and wearing the same thing. The more mature one leads and dominates the play, but is the weaker physically. They like school. This is their first year, and both seem to be getting along splendidly. Did you read the article on twins in "Good House Keeping" this past year? It is very good and true.

MJS: Although the birth weights for the premature twins were listed at six and seven pounds, this was after being fully dressed. The precision of the garden scales left much to be questioned. The doctor was puzzled after the first birth and speculated that a tumor remained. Three hours later, a second baby put in an appearance in the home-birth delivery.

Recently we had Missionary Society for an all day meeting, and I served date roll for dessert. It is easy to make, easy to digest, and a great hit at this house. Date roll:
Mix 1# graham cracker crumbs, 1# chopped dates, 1# chopped marshmallows (use scissors dipped in hot water for chopping), 1 cup nutmeats, and moisten all this with cream to form a roll. Roll in few extra crumbs and chill in the refrigerator. Slice and serve with whipped cream.
 I don't make expensive, hard-to-bake fruitcake. I just make this spice cake, sometimes six weeks before Christmas. Be sure to ice well before storing away—use caramel icing. It makes a big cake. I've made it every Christmas for ten years.
 2 cups sugar; 2/3 cup butter; 1 cup nutmeats; 3 cups flour; 5 eggs; 1/2 teaspoon salt; 1 1/2 teaspoon each of cinnamon, allspice, and cloves; 1# box raisins cooked with enough water to have 1 cup of raisin water to moisten batter; 1 teaspoon soda. 1. Cream butter and sugar. 2. Add beaten eggs, flour, nut meats. 3. Add raisins. 4. Last, add raisin water to which has been added 1 teaspoon of soda. Bake in moderate oven just about usual time for ordinary cake.

March 5, 1942

I just finished supper dishes, helped Martha with her piano (Why won't she practice unless I'm sitting there too?), put Ruth to bed, and now I'll settle down in the quiet and get this letter along. Almost quiet except an occasional "cheep" from the baby chicks. No, I'm not in the chicken house—they are in mine. I have 250 Rhode Islands Reds in their boxes on the couch! Just came from the hatchery and we are keeping them in here tonight until we are sure the brooder stove is regulated. I've planted some "Victory" garden too, but my part seems so small compared to what the boys of our army are doing. It's time we Americans talk less and do more, eh? It seems difficult to believe it could happen to us, but it has.

MJS: My father was deferred from the draft so that he could farm. Farmers were called upon to support the war effort by providing food and clothing not just for our soldiers but for the Allied Nations fighting with us. My folks were very conscientious. We did everything we could to conserve, to raise a victory garden, and to raise extra chickens, hogs, and cattle. Women and children worked in the harvest picking cotton while the older men helped dad with the heavy work. Mother saved our fat drippings and took them to a distribution center for use in making explosives. We had ration books and received only a limited amount of sugar. We did not go places, for we conserved our gas to run the tractors. It was a time of hard work and sacrifice.

Ex-Steno, enjoyed hearing about the babies. My twins are growing so tall now. They are in second grade. Mary has bad tonsils and has missed at least seven weeks of school from bronchitis, flu, and infected eyes. She gets something new every few days. Today I took her to the doctor as she had hives so her eyes were almost swelled shut. She is al-

lergic, but to what? I've spent part of everyday teaching her so she wouldn't get behind her sister. Mary is an average, fun-loving youngster. Martha is a serious, mature, bookish student—so I had to keep Mary busy or else. The teacher was sick, and as teachers are getting scarce, I substituted a few days at school for her. It was sort of fun as it recalled my "youth."

Alice Jeske, golly I wish I had you here. We decided to install a regular bathroom to replace the old system of carrying rainwater from the kitchen pump to the tub. Along with the bathroom, we decided to move a closet and change the bathroom. So a wall must be moved, window put in, and doors changed. Sawdust is tracked everywhere. Sometimes I wonder if old houses are worth remodeling, except in the satisfaction it gives you in believing you are making something out of nothing. The more you do, the more you need to do; so there is no end.

MJS: Taking a bath meant pumping the rainwater from a holding tank under the utility room, heating it on the wood burning stove, and carrying it to the bathtub. Mary and I usually had our baths in the kitchen sink with a teakettle of hot water added to the pumped water. Because we lived in the Mississippi bottom land, a pump could be driven down anywhere a few feet to a bountiful supply of water. The pump installed in our kitchen was placed in a position so that water could be pumped directly into our sink like a water faucet.

We have very hard water and high in iron. We are getting an iron remover—Refinite Company of Omaha. Wonder if any of you have had any experience with such?

Someone got the idea I had hardwood floors. Oh, no, just the old pine floors sanded off and given three coats of clear varnish. They are surely easier to keep than they were when they were black with old varnish. I like to refinish furniture. I removed the varnish from my walnut piano a few years ago and refinished it. *Last fall I refinished two second-hand colonial rockers and use them in my living room. The money went to fix the old house, so I had to fix the old furniture myself. I use varnish remover from Sears, more economical. I use a brush and just slowly wash off the old varnish or enamel. I usually blister my hands, but the satisfaction is great and the saving more.*

Dear friends, teach your boys and girls safety rules on the highways. A twelve-year old lad dashed into the side of our car and was killed instantly last October. We did our best to get out of his way, and it was only a miracle that we did not roll into a ditch. A car was tossing out bills, and he dashed across the road without looking our way. We were all in the car enroute to Memphis. You never know the pain that comes from having been a part in such an accident. Although we were not to blame, as we could not get out of his way, we were deeply grieved. Those were the darkest days I have ever known. It would perhaps have been better if we had been driving faster. We might have gotten past him. Perhaps I shouldn't mention this, but it is often on my mind. I would urge all parents to help their children observe safety on highways. I've sent mine through the pasture most of this winter to give me peace of mind.

MJS: The young lad was twelve years old but in first grade. These were days before special education. My father swerved to avoid hitting him and almost lost control of the car. We bounced around inside the car, for this was before the days of seat belts. The boy was concentrating on securing a handbill handed out by the car ahead of us advertising movies and did not think to look. He was one of three children waiting for a school bus. My folks were devastated. My father said that he didn't think he could ever get up in church and speak again. The patrol officer reminded him, "It wasn't your fault." We could get to our country school by going through the muddy farm pasture and bypassing the gravel road to the school. This gave my mother peace of mind.

Our winter was the coldest we have ever had—20 degrees below. It killed two cedar dedora trees that were in front of the house. They dug them up today and put out some that are more hardy. So many people lost their southern evergreens. We got ours from the Piggott Nursery in Arkansas. That nursery lost $5000 in evergreens.

January 9, 1943

To see all this cotton gives Mary a headache.

Glad to have you come again. It has been a busy year for us, especially the harvest time, and part of the corn and cotton is still in the field. So many of our neighbors and help are moving to St. Louis to work in defense work. I did my part by helping with the cotton weighing last fall. I have a pair of slacks so I could climb up on the load of cotton if necessary, and I weighed in the afternoon while Ralph fed the livestock. I'd never thought I'd wear slacks but find I like them for gardening, washing etc. I ordered them from Lane Bryant, and sure they are big else I'd never get in them!! I figured all the cotton weighing and paid the pickers. Then my mother was sick, Ralph's mother came from Chicago to visit us, the tomatoes were *ripe, Ralph had to combine lespedeza seed, so we hired a girl to weigh. Had to pay her $5.00 per day. The pickers made good wages, some $10.00 per day. Several ten-year old children made $2.50 to $3.00. You see, schools start in August and then are dismissed for six weeks for cotton picking vacation. It was lovely picking weather, not hot, nor cold, and not a rain for two months. The pickers did so well they haven't cared to help get the corn out. They work about two days a week. One family (father, mother, three children under twelve) made $100 the best week. I don't see how farmers are going to compete with city defense plants as to wages, but somehow we must produce for our task is small compared to the men in service.*

I am an only child so I have no brothers for service, and Ralph's brothers haven't had to go yet, but my sympathy goes to those with loved ones in service. My best friend (Virginia Twitty, a Home Demonstration Agent) has three brothers in service. Her mother is so brave, but so disturbed. The burden is heavy when we give three loved ones.

MJS: One of the three Twitty boys was declared missing in action and never found. We went to their house, and my mother talked and cried with his mother, Mrs. Twitty. The reality of war was closer to home.

Today our new church was dedicated. After the service we had dinner in the recreation room. So glad to have the debt lifted now and so happy to have the privilege of worshipping in a land of freedom. It is not an expensive building but adequate.

My hens are doing well. I get 73-77 eggs per day from 110 hens. We built a new Illinois type chicken house, and it has really paid. I got 50 cents for a case of eggs Christmas week, but most of the time I get 40 cents. See I'm keeping books on them to see if they will pay for the new house. Ralph has a switch by the bed to turn the lights on in the chicken house at 4:00 A.M. It will soon be time for baby chicks. We get ours the first of March.

MJS: The strategy was to increase the yield by arousing

the chickens at 4:00 a.m. so they will have more time to lay eggs. The light switch meant Dad did not have to get out of bed to turn on the light.

Last October we visited Ralph's brothers at Urbana and attended the Illinois-Notre Dame football game. It was the first game in ten years for us. But I didn't find it nearly as exciting as I did ten years ago! Did Ralph enjoy it, aside from the fact Illinois lost.

My deepest sympathy goes to you who have suffered the loss of a loved one. May your faith in God's plan of eternity make your burden lighter.

July 1943

Now we are having a little breathing spell, and school will start in another week. Schools start early, and then have six weeks vacation for cotton pickin'. Since a vacation is out for us—and a book doesn't go too far—we are having a little vacation at home by doing a few things we have been too busy for. We go to the river swimming and fishing about one time per week and find we have good swimming and fish real near. Why have we looked for "greener pastures" for vacations? Ralph is talking of going to northern Mississippi for a young bull. If he does, the girls and I will go with him to Memphis and visit the zoo, and if he purchases a bull, we will return on the train, and the bull will ride in the truck. Just a few things like that will give them memories of the summer vacation for the winter months when we stay indoors, and when the fall harvest starts, we will be too busy for such.

There is a pleasure in doing a job well, and I think physical work keeps people happy and well. We farm folks haven't time for brooding and nerves just now. Maybe sometimes we could do with more rest though. My dad is such a worker. He doesn't know how to have fun. I've wondered if we live to work or work to live!

In addition to canning, gardening etc., I've found time for a little "hobby riding." I have just cleaned an old walnut dresser for the twins' room. It has a lovely mirror—about two and half by three feet—in perfect condition. So often the mirrors are bad on the old dressers. It has a perfect marble top, but this I want to change for walnut when I can, and it had a great carved board like a rooster comb on top which I had removed as it was too tall for the base of the dresser. It is low enough to serve as a vanity as you can see your dress hem in the mirror. It had only one coat of varnish so was easy cleaned, but it was just black with dust and dirt. I like to clean old furniture—do it all myself with varnish remover and then oil or wax the piece.

I'd better sign off and send this on its way. Do any of you have relatives at the Malden Airport? If so, I'd like to have them visit us. We visited the field on the Fourth of July, which was open day. Let's pray as we work for peace.

MJS: An airbase was built at Malden, Missouri, to train pilots during the war. Part of the instruction involved the use of gliders. An accident occurred in the summer of 1943 when a glider broke lose from a plane and crashed, killing the two occupants. The accident was about four miles from our home. All the planes and pilots on the base were ordered up in the air, and they buzzed the surrounding area to keep the minds of the young instructors and their students occupied. We stood quietly in our yard and watched.

January 20, 1944

We have just had a loud session of radio. The twins are just beginning to listen to radio. Tonight was their favorite, "Snooks" and the "Henry Aldrich" program. They jogged some chuckles out of me too, even if I was rather tired. "Henry" can get in the "darndest" places! In this time of stress and hard work maybe we need more such humor. I don't care for any of the war movies, but for an evening of laughs do see "Hit the Ice" (Abbott and Costello) if you have a chance, or for beauty see "My Friend Flicka." We rarely go, but if we do, we want nature's beauty or humor.

I've canned thirty pints of fresh pork this afternoon. In

fact I have a new cooker now. It is my first time to use one of the new Victory models. It has weights instead of a gauge, so I don't have to watch it every minute. I don't like butchering, but we do like the canned meat better than any we buy. When I cut up the meat, I lose my appetite for it for the time. Maybe that will save me an added pound.

That's Seco Lo Down, whose uncle sold for $4,000.

I've been making a seed order. I'm going to try Burpee's Tampala, a new vegetable. Are any of you trying?

So "Happy Rancher" will be in Texas again. I know there is a lot of work with moving, but it seems "sorta" romantic too—change of scenery. Having lived here twenty-five of my thirty-three years, I sometimes think I'd like to move!

Molly O'Neill, you have lovely daughters. I might be partial to girls—having three little ones. Nice pigs too. Like you, our big money is pigs. Ralph said they were mortgage lifters, and so we raised them. Just as we lifted our debt, the war came and now we feel it is our duty. Now he is getting interested in purebred, so I imagine we will keep on after the war too. But gee, they are messy creatures and don't add much beauty.

Alice Rule, you do get about! The snow sounds interesting. I've always wished two things to do! Play in an orchestra and be able to ice skate. We rarely have ice for skating, and I couldn't if we did, but how I enjoy watching others skate.

June 22, 1944

This Robin arrived the day they started to remodel our house, and of course I'm not done yet, but at least there is more order in the confusion. What did we do? We made our front porch into a living room or sun room with eight windows, added a small roof over the front door, lengthened the dining room four feet, added a large dormer window on the south roof, added a gable to the east so we may have an up-stairs someday, added a large back porch to accommodate the front porch furniture, (the old back porch becomes my work room for washing and ironing), had three floors sanded, had five rooms re-papered, the kitchen repainted, enameled the dark varnished wood in three rooms (ivory), had a chimney built for the new living room, re-built the old chimney part way, had a book case built, and last but not least had the old shingled roof removed and a new fire-proof roof put on. The new roof was the real cause of all this. Ralph said if I contemplated remodeling the house for the next twenty years, I'd better do so when the new roof went on—so I thought up all I could.

Remodeling has been our topic of conversation for a year. Of all the mess and dirt, shingles and shaving, and moving of furniture, I've had it! My curtains in the front of the house had been down for six weeks, so I stayed up late last night and pressed some new mail-order curtains and hung them. They are ruffled tiebacks. I wonder if I'll ever get them in shape after I launder them. The largest size is 208 inches. The house now has a living room, dining room, kitchen, bathroom, workroom, and four bed-rooms counting the one for hired help, which we seldom have, and an unfinished room upstairs. It is fawn colored with brown trim. And of course added to all this, it was time to can peas, asparagus, strawberries, and the last of school activities.

Blossom, I wrote Theo Presser and got several books that are a great help with the piano lesson. My difficulty is finding time for the lessons.

Missouri Shepherdess, I planted some berries, but no luck. Several of our neighbors have magnolia trees. Just now they are blooming. The blossoms are the size of a pint cup, a cream white. Both the leaves and blossoms look as if they had been made of wax, and the flower has a delicious lemony odor. They always recall that I was a June bride, for there was a bowl of magnolia blossoms on the table at my announcement party. June the month of brides! No, I'm not getting romantic. Our youngish minister just married one of our local teachers. It was rather logical if marriages ever are? She will make a good wife for a minister. It must have pleased the congregation, for they have certainly received a bountiful shower of gifts, even a $50 wool living room rug. This was quite a lot for a small congregation.

The above paragraph contains everything from flowers to preachers! Maybe that is a sign I better cease this rattle and off to bed.

July 2, 1944

I attended Extension Club until late one night and then the following night a dinner for 4-H Club Council. Our club members are all working in stores, post offices, or offices as their husbands are called to service. They take their job or run their business, so we meet at night as that is their only time off. Our lesson this time was all the latest on canning. I don't see when they will can, but they say they will. For me, I've canned kraut, green beans, diced beets, and cucumber pickles.

I hope it continues cool for a few days as we are threshing wheat. They have combined some, but most of it was cut with the binder, as we like the straw stack for the herd. . . .

MJS: Cutting wheat with a binder was the old-fashion way of harvesting wheat. A man would drive a mule to pull a binder which would automatically cut the wheat and tie a string around a bundle of it called a sheaf. Then the men would walk around the field picking up all the sheaves and stand them up in piles with the heads of the wheat leaning against each other. These piles were called shocks, and the task was called shocking the wheat. I loved to watch the men comb the fields and set up the shocks. It looked like play to me, but it was hot, hard work. To see the finished product was astounding—a vast field full of golden mounds of wheat. Then the hardest work began. The threshing machine was pulled into the field and threshing began.

The shocks were carried by mule drawn wagons and fed into the threshing machine. The threshed wheat poured into sacks, and the straw was blown out a pipe that made a mound of golden straw called a straw stack. And, as mother said, the straw was used to feed the herd. Soon combines took over the chore of harvesting crops, first a two row affair, then four, six, and now eight row combines. We no longer have straw piles.

. . . The past three weeks have been a mad rush—the cotton to chop and plow, the corn to cultivate, wheat to cut, and alfalfa to cut. Children and women chopped the cotton—cost $3 for a ten year old, $3.75 for older.

Last week Ralph brought his Sunday school boys out for a swim in the river near here, and we had them for supper. Next week I plan to have my class of senior high girls. One girl raised an objection to the fact some "Italian prisoners—farm laborers" were also going to this same place to swim. I kept quiet and listened to the different opinions. The racial problem is always coming up in my class. We have lots of colored or Negro people here, and they do create a problem. The North and South are miles apart in their views.

MJS: A prisoner-of-war camp for Italian soldiers was set up at Marston, Missouri, a town six miles from the farm. The soldiers were used as farm laborers, although they did not work on our farm. It was a rather relaxed prison with low security fencing around the compound. These were some of the early prisoners in World War II. The girl in Mother's Sunday School class resented swimming with the Italian prisoners, just as the girl would have objected to swimming with African Americans. My father who was from the North could not understand the views and prejudice of the South. He was appalled to learn that the only high school for black students in the county was on the third floor of the New Madrid Court House.

We are enjoying a nice supply of garden vegetables now. I believe our favorite meat is fried chicken, vegetable is tomatoes, and dessert is freezer homemade ice cream. The ice cream has cut a great hole in our present sugar supply, so we are leaving it off now.

The twins are having their first 4-H work, and it has surely stimulated their interest in cooking. Since Ralph and I are both leaders, Alma Ruth, age 5, goes to most of the meetings. She always repeats the pledge and joins in songs and games. "I'll be all ready when I'm ten years old," she remarks.

I have always sewed for my girls so have got along very well on dressing the children in wartime, but now it is getting to be a problem to find materials. How do the rest of you get along? One store had two pieces of cotton material. I'm not meaning to be a complainer. I just wondered if the shortage was just here because everyone wears cotton in the south so all the "prints" had been sold.

MJS: For 1945 I have no letter. It was the year Marjorie Alice was born and the war ended, a memorable year. We had a gully-washer in June which left the farm a lake. Dad brought a boat up to the back steps of the house and let us paddle around the pasture between our house and our grandparents' home while Mother did hand sewing and watched our Huck Finn escapades. Later in the week when Mother's water broke, Dad took her to Southeast Missouri Hospital in Cape Girardeau. We stayed with Memo and Granddad (Mattie and Joe Heath) and went each day to the closest phone at Catron for news. The first day, no news. The second day Memo became concerned. Finally, the third day, we learned we had a little sister who was born on Ruth's sixth birthday. The nurse was amused because the first question that Dad asked was, "Does the baby have black hair?" That was more important than whether the baby was a girl or boy. Because Dad was one of five boys, he treasured having girls and made us feel very special. When Mother arrived home, she put Alice on the bed in the front bedroom, and we surrounded her with our chairs and watched her all afternoon. Mother had written that she was a big baby. She looked so tiny to us but fascinated us with her fluttery hands and her kitten-like squeaks.

August 12, 1946

Much has happened since you were here. We have four daughters now—Martha and Mary (twins), age 12; Alma Ruth, age 7 (June 16); and Marjorie Alice, age 1 (June 16). Two birthday parties for four! Saves on sugar. Alice is a very lively little lady, walks all day long, talks a little. She is a great joy. My babies are six years apart so I forget so many little things that a little one recalls again. Alice was to have been my son--so was Ruth--but by now I have forgotten my disappointment. Sons are not for me, I guess.

Washed today and mowed the "back yard" which is enclosed with white pickets and is our favorite place and Alice's "camping ground." Ralph eats with the Kiwanis Club on Monday evening, so the girls and I enjoyed a cold watermelon and didn't have dishes to wash. Now while I write, the girls are reading "Huck Finn" aloud for Ruth's benefit. We read lots of good books aloud, have just finished "Lassie Comes Home." I usually read three chapters aloud each day. Ruth is such a good listener that we enjoy it even if she is young. I surely recommend reading aloud for family pleasure.

MJS: Many winter evenings Mother recycled the book list from her teaching days and we got the full treatment -- "Little Women," "Little Men," "Ivanhoe," "Black Beauty," "Evangeline," "Rhyme of the Ancient Mariner," "Little Lord Fauntleroy," and poetry of all dimensions. I well remember "Girl of the Limberlost," because the hero dismounts from his horse, removes a leaf covering the footprint of his ladylove, and kisses the footprint. I thought it was terribly romantic, but from Mother, Daddy got, "Now, Ralph, no smirking." Ruth was wonderful to read to because she cried in all the appropriate places.

To be quite frank, I've spent the summer worrying over Mary. She had rheumatism in her feet and ran a temperature. Dr. said rheumatic fever and wanted her in bed all the time. She felt better, and it was hard to keep her quiet. Then someone would come along and say I should by-all-means have her in bed. Called another doctor. He was puzzled and not sure, so he recommended we take her to Children's Hospital in St. Louis for diagnosis. This we did. She only had rheumatism from very, very poor bone structure in her feet. Was I ever relieved after three months of worry! Now she takes foot exercises regularly and can go to school. The twins ride a bus ten miles this year to school. Ruth goes to the rural school. Ruth and I stayed in St. Louis while Mary was in the hospital. We visited the zoo and Shaw's Botanical Garden.

MJS: Rheumatism was a catchall diagnosis when doctors did not come up with anything concrete, something akin to arthritis. Mary spent the summer rolling her feet on a rolling pin to improve the muscle tone. In later life a doctor confirmed that she had had a hip break sometime in her life. This break would account for her crawling to avoid the pain of walking.

November 2, 1946

Monday night my 4-H girls had their local achievement day. Wednesday night the Youth Fellowship had a Halloween party. I led the games and made costumes for my three girls out of next-to-nothing. Thursday night I was chauffeur for another Halloween party. Friday, I went to the locker plant as we were out of meat, and today I've cleaned, pressed six wool skirts, made head cheese and made kitchen curtains (found some permanent finished organdy for 49 cents).

My pet peeve just now is the shortage of material for children's' clothes. I just don't have time to drive miles hunting slip material. Where does all the cotton go? Now I am told Martha Jean will represent her room in the queen contest in the Annual Variety Show, and she is to wear a formal. Where do they expect me to get such? A twelve year-old girl has little need for a formal. I'm not going to spend lots of money for one, and I can't find material to make one. Monday morning I am going in search of something; maybe some curtain material might work.

Marjorie Alice is the "nosiest" girl I've had. She is very energetic and constantly prowls and pulls things. She is such a great pleasure, but my house is always littered with toys, magazines, dishes, etc. Usually at least one pitcher is setting in the floor. She can reach the pitchers.

We attended the family reunion at Urbana and saw Notre Dame beat Illinois. It was good seeing all the little nieces and nephews. There are thirteen grandchildren including our four girls. However, we didn't take our baby girl as it was too hard a trip for her. We visited all the Polled Hereford herds in southern Illinois, but didn't find just what fit our purse until later.

Have you tried giving yourself a home cold-wave permanent? My neighbor gave me one, and I gave the twins one. It is the nicest, most natural wave I've ever had. Had it cut first by a beauty shop operator.

September 8, 1946

We made a flying trip to the fair, visited Lincoln's home and tomb, the Lincoln shrine of New Salem and spent one night in Peoria, Illinois (visited Ralph's brother). We loaned Alice to my mother, as the trip was too hurried. We had fun. Especially did I enjoy New Salem, reconstructed as the village was when Lincoln lived there. I've been thinking of the antiques ever since! We were in a hurry because our girls are in school—have been since July 14. They go to school at 7:00 a.m. and are home at 1:00 p.m. By leaving at noon they were not absent. They have another week of school and then they will be out for cotton vacation two months—sounds sort of silly.

MJS: The school bus came early in the morning so classes could start at 7:00 a.m. School was dismissed at 1:00 o'clock because of the unbearable heat. None of the school rooms were air-conditioned. Classes dismissed for cotton picking vacation as soon as cotton was ready to pick in August and September.

This has been such a warm summer. It is 100 in the shade on my back porch now. I'm perspiring from every pore! We haven't had rain in so long crops and pastures are drying up. Cotton likes dry, hot weather, but it could stand some moisture now. *The flowers have all quit blooming except the gallant marigolds.*

We are trying to stretch this house again and are adding a stairway to an attic playroom, three new hardwood floors, some new closets, a bay window with a picture window, and a few other things. We tried for six months to get a carpenter, gave up and hired a large contractor. He sends his men a few days and then on another job, so we are making little progress. In the meantime Ralph built a large implement shed--80 by 40. He says that is more his line. He is afraid to undertake the house. I just wish the contractor had Ralph's ability to speed this job along.

I have been quite busy with 4-H club activities this summer. I even went to camp for three days—Camp Arcadia. It's a lovely spot in the Ozarks, except it was hot there too. I even climbed Mount Pilot Knob. Did I ever puff! Came home resolved to get rid of some of this excess I tote around. I am losing a little. If you could see me just now you would declare I'm melting.

MJS: Pilot Knob is a mountain with a scenic view. It was at Pilot Knob that one of the battles of the Civil War took place. Camp Arcadia was a well-established Methodist Church camp near the mountain.

April 27, 1947

It is spring at last! We are having asparagus, onions, and radishes from the garden. The irises are beginning to bloom. I have several different colors and some new ones I am watching. My greatest pleasure is working in the yard and garden. Some women move their furniture for variety, and I move my flowers! The yard near the foundation of the house was hauled in and not the best so has caused some changes. Tomorrow I must prune the evergreens, should have earlier. I have some nice ones, and how they grow! Have been putting a privacy hedge between the garden and me and set a dozen new tea roses. Ralph bought a new power lawn mower. Maybe I can lose a few pounds trotting around after it. It really travels.

Mandy, we are twelve miles west of New Madrid on the Mississippi River. You can stand on the river levee and see the states of Kentucky and Tennessee across the river from Missouri. The Missouri "boot-heel" as this part of the state is called is the little southeast corner joggling down into Arkansas. Lilbourn is small—about 1000. We are on Highway 62, five miles from Highway 61, a direct route from St. Louis to Memphis. Come by.

Elora, your house and antiques are so interesting that I read your letter again tonight. New Madrid is one of the oldest towns in the state. Several of the old homes have lovely antiques that are family heirlooms. A great part of our feelings for antiques is the sentiment of course. One year the Farm Bureau had a tour through one of these homes. There were two of the large mirrors standing on legs with much gilt. The wallpaper in the parlor had been hung before the Civil War and still was in fair condition. However, in spite of tradition, I believe I would have had it papered. I would have tired of the huge gilded flowers. I agree with Mandy that pieces should be retained as they were originally to be true antiques, but in my case I bought a chest, and later a walnut dresser in a second-hand store, cleaned them, added square walnut knobs, and used them along with my other furniture—some modern, some period. The chest, as I used it, had three tear-drop handles or pulls. Where would I ever find others? The dresser had what appears to be copper hardware type pulls. I bought it first for the mirror. The mirror is about 42 by 36 inches and in perfect condition.

Molly, some of the neighbors insist that some of the mums mix. In other words, if you get a certain bronze mum, then the following year most all will be bronze. Didn't seem logical to me, but I don't know much about mums. Would that be true?

I made myself a jacket dress Thursday with lots of fuss and worry. Ralph teased me about it, said why didn't I buy the dress. I insisted I was too hard to fit. He insinuated I was hard to please! As a result I went shopping in earnest on Friday, and to my amazement found, not one but three, that would do. Now he is sure he was right, and I have more new dresses than I have had since "before the war."

Our Marjorie Alice is growing up so fast. She tears around outside, swinging and playing with the kittens. She says all the words and parts of sentences. Ruth is trying to teach her to count. I believe she is rushing her. Alice will be two June 16th.

January 15, 1948

This round of letters is especially interesting, and the pictures so good. We have had difficulty getting film for the still camera, so most of our recent pictures are movies. Then I realized I didn't have any for Alice's baby book, so my aunt in the city sent us film. We really enjoy our little movie camera; the colored pictures of baby Alice are lovely.

Ilene, in regard to you personally, we were surely listening to that Rose Bowl game! Ralph loves sports of any sort. He is not heavy, but rather athletic. Tonight he is attending a boxing tournament at school. He tried to get the girls interested, but they were definitely not interested. Sorta hard on a father to have a house full of women folks, eh?

Yes, after visiting three small towns, I found delicate pink cotton net; and, luckily, enough satin for a slip at 79 cents per yard. I made a simple bodice dress, tied it at the waistline with a velvet ribbon and let it hang to the hem. She (Jeanie) looked lovely to her mom! Your twins, like mine, are quite different. Martha is musical, athletic, and "bookish," so it makes a difficult comparison for Mary Ellen, as she is in the average class. But fortunately, Mary is a sociable, happy disposition, and hasn't bothered or worried that the honors fall more often to Martha.

Mandy, hard to say which was most interesting, Gini or the antiques—I have a weakness for both. I only have a few dishes (a blue pineapple butter dish, the lid is shaped like a bell and did have a copper clapper, a blue hob-nail cake plate, a jelly dish, and an old shaving mug). I have a lovely walnut dresser and a chest I bought cheap in a furniture store.

I had to stop to set Alice on the "Little" chair. That's what I've spent most of January doing. Seems about time that gal took a hint about what panties are for! My newest sister-in-law with her first baby (two months younger than Alice) was here in December. She sat the baby on the "Little white" chair every few minutes all day, and with little success. I'm sure she thought my baby methods very "antiquated," but I was too busy with the rest of my family to waste too much time until I was sure of some progress.

May 2, 1948

You always arrive at a very busy time, but it seems all my days are usually busy ones. This is our last week of school—the girls have invited girls to spend the night Monday; there is the annual musical program. Also I must find time to have a 4-H project lesson, attend "Guild" and finish two dresses, for the twins finish eighth grade this year. Then, as soon as school is over, we have two weeks of vacation Bible school, and I promised to teach a class. Don't we Americans sometimes rush around too madly?

We did finally get through the remodeling. We do enjoy the improved space. Now my problem is to find a few new pieces of furniture. Have just received a new Olson rug and am ordering another like it for the dining room. The Olson rugs surely wear like iron, but are hard to get as you have to wait to get on their list.

This is the season I like best. We have asparagus, onions, radishes, lettuce, and in a few days, peas. (I usually have peas before Mother's Day.) Flowers are blooming everywhere—peonies (three colors), iris (three yellows, white, about six shades of purple or blue), and lots of roses. We set out twenty Paul's Scarlet roses on the front fence. I'm anxious for them to grow. The air is balmy; there was a much-needed rain last night. Now farmers will be busy finishing planting. Some cotton is up. The ladino clover is blooming, ready for the first cutting. There are baby calves and pigs. Tippy (Persian mama cat) has four husky, happy kittens, so it is surely spring at this farm.

Blossom, the poem is dandy, so I saved a copy. You mentioned the cold. We had more snow than usual this winter too.

My teakettle is used for pump priming. (We get water by driving a pump just anywhere in this section. These freeze sometimes in winter, so the men carry a teakettle of hot water to thaw them.)

Molly, I know you will enjoy electricity. I didn't think we needed a deep freeze. Ralph is mechanical and gadget-minded, so he found a small one that just suited him. Now we have one in the workroom. We also have a locker at town, and they do the butchering. Both are full, and I keep a better supply at home and am always prepared for company. We even keep spare loaves of bread. They are just as fresh as when we put them there.

Mandy, I have a 15th wedding anniversary June 4. I want a table that unfolds to seat twelve and the rest of the time occupies a small corner. I really have two living rooms instead of the usual dining room furniture.

Alice, I'm still loaded with 4-H'ers too. We have "Yard Improvement," "Home Service," "Cooking I," "Vegetable Gardening," "Cotton," "Fat Barrow" projects in a club of twenty-two. Mary won a blue ribbon on her hog. He weighs 306# at six months, brought $84 in the sale. Martha has the baby chickens as her project. She was sorry she didn't have a hog when the Annual Show time came. The pig was named "Oogie," and of course was quite a pet.

MJS: Show time for Oogie took on the hoopla of a county fair. A large tent with a show ring was set up in New Madrid for the Fat Barrow Show. From all over the county 4-H'ers brought their pigs to be housed in display pens. Pigs were washed, combed and paraded around the ring amid much squealing and prodding. Contestants were proud of the price their pigs brought but teary-eyed to lose their pets. Judging and sales were handled according to weight classes.

The project to care for chicks and keep a tally of their productivity was my 4-H project. A baby chicken house was built on wooden runners while the adult chicken house was built on a concrete slab. I fed the baby chicks until they were old enough to move into the big chicken house with windows across the front, a roosting rack across the back, nests on the side and a holding coop for the chickens who insisted on "setting." As baby chicks got older, their house was pulled with a tractor closer to the adult house so that the chickens could roost with the full grown chickens. We were interested in the egg production and discouraged chickens that sought to set on their eggs. The setting hens were placed in a holding coop with a wire floor so they could not nest. My project was to feed, water and keep a tally of their egg production.

Elora, I have a vigorous bittersweet vine, but don't have any berries ever. I suppose I need two. Your berries sound lovely. Your work with sheep sounds interesting. Ralph just returned from church (night service), and has gone to the barn to care for a new calf. He enjoys animals, but they can be lots of work.

November 11, 1948
We have a high, cold wind tonight following a heavy rain, so I suppose this means the end to a very lovely autumn. We had a very good crop season and had two months of nice harvest weather. School was dismissed seven weeks for cotton picking, and many bales were harvested; but as usual there is cotton to be picked. The girls were proud of the money they earned to buy most of their school clothes. Prices were so good that many people—school teachers included—picked who had not ever picked before.

This was the busiest summer ever. In August the Balanced Farming Day was held here on our farm—400 people came to see how we farm and live. Then in October farm and home agents from Mississippi came to Missouri to observe the Balanced Farming Program. Each visit made it necessary to do some extra house cleaning, so I was always busy.

My twin daughters are in the first year of high school and most as tall as their mother. They surely liven things up—I can scarcely concentrate while one is practicing a saxophone, the other the piano. They gave a demonstration on "Three Arrangements of Annual Flowers" this year. You can be sure I had to learn much about flower arrangements, and I am sure it was good for all of us.

Mary wears braces on her teeth, and so we must visit the orthodontist once each month in Cape Girardeau, sixty miles from us. She has been going about three years, and I hope will soon be finished. Of course I make it a shopping tour. It has been expensive to have Mary's teeth straightened, but I feel it has been worthwhile to her health and appearance, as she was becoming a "mouth breather." There are five children from our community going, perhaps twenty from our county. One child's speech has improved as his teeth are straightened.

February 21, 1949
Last week I saw eleven robins at one time in the back yard. The jonquils are blooming and also japonica, so maybe spring is not far off. Yesterday Alice began to break out with chicken pox. She isn't sick, keeps playing, but is bothered with them itching and keeps me busy applying calamine lotion. This is the first time she has had any of the usual childhood diseases—but the schools here have been having everything—mumps, measles.

Last month in our Extension Club we made etched aluminum trays. It took almost a whole day, lots of work and mess, but we had a lovely twenty-two inch tray for $1.00. Everyone was thrilled with her finished tray, so we are planning to make more. They look like silver, use a pattern of your own choice; "loveliest" were the ones with a monogram.

I am refinishing a picture frame. It is the type with the scrollwork in something like plaster and gold band next to the glass. Now I shall gilt it all, then add a coat of white flat paint. When the paint is nearly dry, I sand the flowers lightly to let the gilt show in high places. It is a large picture frame, so now I must decide on a subject. I have written to some companies but haven't found what I want. So much of modern art disgusts us, or we fail to appreciate it. I bought it for $1.50 at a sale. Framed pictures like it range from $35 to $65. My family "kids" me lots about it, but they are interested too.

Ilene, we saw and read so many stories of the trouble the snow caused the western ranchers. Made us a little smug that we didn't have such weather. Aren't humans like that? I'll bet you are glad that when it melts it won't cover you up. We humans build us a nest and then stay there in spite of its handicaps. This is home to me since I was eight, but I haven't lost hope that someday I can see how some of the other Americans live. I know if I ever build a new home, I am going to hunt a hill first. I know Ralph's reply is always that the soil on a hill is not worth our fields.

Alice Jeske kept the pictures of Martha and Mary by error and then mailed them on to me. The girls were fifteen on February 12. They are tall as their mother now. Both are blond, especially Mary (Her hair is quite golden, and she has a nice complexion. She looks dark in this little picture though.) They are in first year of high school and very busy with school, club, and church. Both play in the high school band, and Martha has played piano for the church service the last two Sundays. She has played piano for the children's department since she was eleven, but our choir director (also church pianist) was needed to sing, so she called on Martha. She was rather nervous, as well as her mom the first Sunday as there were two music teachers in the choir, and the director has an excellent musical education, but she didn't make too many errors.

Ralph is busy making the attic into a playroom. All the girls around here love to play in the Hubbard attic. Of course it has pieces of all the toys from the twin's day up to the present. Right now it is the grandest mess of toys and sawdust, old books, and boxes you ever saw. As soon as he finishes, my job begins!

June 5, 1949

I have just finished two weeks as a vacation Bible schoolteacher. Our church is ten miles away so this filled my afternoon. I taught the junior high group. For study each made a miniature bookcase filled with books of the Bible. The material came in assembled packages and required study and copy work to assemble. For recreation we did textile painting on feed sacks, also used old white shirts to make aprons and painted these.

Just cut the shirt into under the arms, add a band and ties, move the pocket and paint across bottom. We used white shirts with worn out collars and made pretty white aprons. They surely did enjoy this and had one messy time!

Alice and Ruth pose in front of the big ditch on their shared birthday.

The twins are away from their family for the first time tonight. They went with a group of 4-H'ers to Columbia, Missouri to a camp. They were quite excited. I can see I'm not looking forward to having them finish high school and going away!

On June 16, Ruth will celebrate her tenth birthday and Alice her fourth, so I must plan a party. Alice is growing up too fast. She tries so to be like the big kids. We have a flute for Ruth. She doesn't know about it and hasn't remembered that we gave the twins a horn on their tenth birthday.

I hear a mocking bird singing away in the apple tree. The storm blew down so many nests, and Ruth rescued a red bird. We kept it a week, fed it baby chick mash after we read that red birds feed on seeds. It was clever, cheep for food, and fly on our finger. Then someone let the mama cat in, and she ate bird! She has been in disgrace since.

November 1949

As you see we visited Memphis. Ralph attended the National Polled Hereford sale; the girls, Mom and I shopped and visited the zoo. A visit to Memphis over night at the Gayoso

Hotel is my daughters' idea of fun. It is all so different from our life. Like their mom, they love country living, but they like a city trip.

Edna, I can sympathize with the lumbago. For one week I could not even turn in bed. This was followed by three years of spells of lumbago. I took big pills, little ones, pink, green and yellow ones, various liquids. In desperation I tried an osteopath. He said I had an injured vertebra. He gave me a treatment and now I seldom have lumbago. At one time I thought I would be stooped as a result. The medical men are angry if you mention an osteopath, but I'm sure they have their place.

Alice, when Ralph has asthma, we talk of Texas or Arizona. Does your son have it there? Ralph uses Adreno-Mist from S-K laboratories in Phoenix, Arizona for relief, but he is having trouble most every night.

A cousin's daughter married last week, and my twins were the bridesmaids. It was a little church, and I decorated the altar with my mums—large baskets and candles. It was truly lovely, even if the work of amateurs. It always seems such a waste to buy flowers from a florist when we have loads of them. Martha truly does nice work sewing now. She made her and Mary a dress for the wedding. Mary enjoys cooking more, and beats Martha there.

Ralph and I had a two-day trip and attended the Illinois-Michigan football game as did 71,000 other fans. I don't care so much for the game, but I love the bands, the color, the crowd. Michigan had the fastest stepping marching band I have ever seen. We visited Ralph's brother over night. We didn't take the girls because they have had polio epidemic there, and I was afraid.

Chapter 5.
Matters of the Heart, Alma's Letters (1950-1972)

February 2, 1950
Dear Robin,

What a fortunate time to have you arrive! Always before I was up to my ears in work when you came. Why not this time? We have had an ice storm, and the electricity has been off two days. It's too bad to visit, so I've indulged in reading a book, "Arctic Mood" sent to us by our Alaskan schoolteacher cousin.

Ilene, electricity is the most wonderful hired girl in the world, but don't throw out your old "stand-byes" too far when REA arrives, for sometimes there are storms! Everything I start to do I am hampered, so I gave up by doing the minimum and declaring a vacation. I am getting by with as little water as possible in the kitchen and bathroom, can't iron although the clothes are sprinkled, can't sweep or sew, and not even a radio. Ralph's taste runs to gadgets and his latest purchase was an electric dryer and an ironer. I can see it is grand for men's shirts, jeans, and flat work. We have had two bright days in January, so I have used the drier too. However, I have saved the sheets for some bright day that I hope will come soon. We heat the house with bottled gas, and it is a moist heat and does not dry things quickly.

Edna, I'll bet you—like the rest of us—have been studying the seed catalogues. I think of all I'd like to get and then

The farm shop is flooded by a big rain.

I remember sometimes we do have floods. Two weeks ago our farm was almost a solid lake, and they were talking of letting the water in the New Madrid Spillway (that would mean the river would be getting closer to us). However, the threat passed. I surely wished we had just one small hill!

We have a small deep freeze, but I was worried that it might thaw if the current doesn't come on soon, so decided to use freely of its contents, a large pan of chicken and frozen asparagus. I also brought out the large angel food cake I had frozen in case of company and a big box of strawberries. Most of the cake is gone. The schoolgirls were hungry when they came in, so I agreed that all should have some—now Alice has had more than some. She must be really filled! She is surely happy now. I hear her singing "Mule Train, Clipity, clop, clipity, clop, a dress of calico for a purty Never-Jo." She sings with great gusto when Martha plays it. She carries the melody nicely.

Our crops were not up to par last year due to too much rain. Now the government restriction program on cotton will result in lots of people going elsewhere for work. The result has created a relief need, but it is exaggerated. It disgusts me to see people so eager to chase the government to get something for nothing. One man here on the farm spent several days chasing around trying to get on the relief roll. When he doesn't succeed, he comes out to work, and we pay him $4 per day to cut up dead trees into wood. We give him the wood, yet he comes up and asks me to call the gas company and tell them to bring him a bottle of gas for his new gas stove. It worries me to see Americans so eager to howl for relief and do so little to help themselves.

For several years I haven't found time to read—now this January I have read five books. We now have a county library and a bookmobile.

Martha is on the debate team. The question this year in Missouri is: "Resolved: that the President of the United States should be elected by the direct vote of the people." Pretty interesting discussions go on in this household now. At first I was sure the affirmative was right and didn't see how the negative could make a question out of it. After reading several books on the subject and going to the debate tournament, I am on the negative side. So!

MJS: The topic raised the issue of using the electoral college versus direct vote to put a presidential candidate in office. This was accomplished in the Bush-Gore election in 2000 with

Gore winning the majority of the popular vote, but Bush was elected through the electoral college.

The doc says my blood pressure is up, and I should take off some weight, as my heart is not doing just okay. Reckon I will. I nearly have to starve to lose, but I know it would be wise to do so. Bet if I didn't have to cook for this gang I could leave food alone—why do some of us have such splendid appetites?

November 15, 1950

Yesterday was my birthday, and now I am forty! "Life begins at forty," so they say. So guess I'm all set to start off this day with a bang. We had my birthday dinner Sunday, invited friends at church. Jeanie had baked a large sunshine cake with lemon jelly—my favorite. Gifts? A new—much needed Frigidaire—a bigger one that I hope will hold the milk, eggs etc. plus all the other.

Blossom, we spent part of the week at Columbia. Jeanie had a demonstration "Making Ice Cream in a Hand Freezer," and Mary entered the cherry pie-baking contest. Both won blue ribbons. We had a nice time—attended two shows, football game, stopped at Shaw's garden in St. Louis and saw the Dahlia show. We stop on the way to shop for a spinet piano—I thought—and bought a Baldwin baby grand instead. I've sighed and sighed about the amount of space it occupies, but it really delights the girls and how they do practice! Maybe they are afraid I'll complain about the size again. It does have a lovely tone which is what Jeanie values, and I let her have her way. (It had been used and was reasonably priced, yet in excellent condition.)

Harvest is late here as the whole season is late. The combines have been going through water to get the soybeans. The cotton is very late and a small crop, so you can expect cotton materials to be high another year. Some sold cotton for as high as 42 cents per pound—highest price I can remember.

December 1951

You are due on your way, but my "sugar plum"—Alice, my six year old—has the measles! She was sick eight days before she broke out, and we were not sure it was measles until yesterday. She goes to a large consolidated school and several beginners have had measles. Guess she will come home with all the childhood diseases. Her greatest worry has been that she would miss being the "walking doll" in the Christmas musical program. I made her a blue formal, for she was a flower girl in the Queen Contest at school, and now she was to wear it again.

Ilene, I know what you mean about gadding and joining clubs. They take a lot of time but do add lots of interest. We have just made large glow candles—molded in square paper milk bottles—for Christmas. They are lovely and cost about 25 cents each. Next I want to try to "decoupage" pictures and am gathering wool for a hooked rug. I'm not the least artistic—and I'm busy too—but I do like to learn something new.

Alice is ready for first grade and the measles.

Edna, sorry to hear you're not being well. Do have the blood pressure checked often, for a hemorrhage can be serious. How I have learned. Ralph is so well now it seems like a very bad dream. He has regained his 26 pounds, his vision is normal, and he has resumed all usual activities. Thanksgiving Day he sang in the choir again for the union service. You can know how thankful for everything we were. His hair is almost grown out again, and the scar does not show. I feel God heard our prayers. The experience taught me many things. This world is God's and how little we know what changes the future hold for us; and *there comes a time when we can only pray, "Thy will be done," and, then how dear and what a comfort our friends can be at such a time! The days were so long. I spent from 7:00 A.M. to 11:00 P.M. at the hospital, but there were always cards, phone calls, or flowers. I'm really a country person, and I feel so alone in the city.*

MJS: At age forty-one my father had a cerebral hemorrhage. He had excruciating headaches for two weeks while an aneurysm was seeping blood in the middle of his brain. Then when it broke, he slumped in the floor and was taken to the Sikeston Hospital. Pressure controlled the aneurysm so that doctors could talk to him, had him stand and checked all reflexes, but when a spinal tap revealed blood in the spinal fluid, he was transferred by ambulance to St. Louis. He was put on bed rest, and doctors waited an agonizing two weeks until the blood was absorbed. Then in a three-hour operation, doctors dripped atomic salts in the carotid arteries and x-rayed the brain to see if the aneurysm was in a place that could be accessed. Surgery followed in which the surgeon removed part of the frontal lobe and put seven clips on the artery that was bleeding. These were the days before hospitals were air-conditioned and before intensive care units were set up. Immediately after the operation, dad was reaching for mother's hand saying, "I'm so glad the surgery is over." Then swelling took over, and his temperature shot up to 106 degrees. Mother stayed with dad and hired two nurses around the clock to work with him. Doctors said that he was like a young soldier at the peak of physical fitness and made it through. One eye drooped, and he had double vision for a while. But, according to the doctor, as the optical nerve healed, the vision became normal. He was ecstatic and had a marvelous year of farming.

Blossom, the soft corn sounds like some of ours. "Only fit to feed the hogs," is Ralph's comment. At last I've persuaded him not to raise hogs—after twenty years—so I wonder what he will do if he has soft corn next year. I'm urging him to take life a little easier. Girls aren't much help to a farmer. We need a son! Martha wants to be a doctor, (I hope she changes her mind.) Mary, a nurse. I'm trying not to say too much; after all it is their lives. Alma Ruth says she will major in Home Economics—consolation because the twins aren't interested. Mary is first in Home Economics at school this month; maybe that will stimulate her interest.

Elora, we went "nutting" for hickory nuts. We went in the woods of the spillway, the area they let the Mississippi River in to save Cairo just above New Madrid. It was a lovely afternoon. We got several bushels—there were tons of them. Ralph teased us for we also got poison ivy. And then the hickory nuts are so hard to shell, and pecans are thirty cents a pound here.

MJS: The spillway is an area set aside within the confines of levies in the Little River Drainage System. It is used when the river rises above a certain level to save the arable land from flooding.

Molly, I heard the weather reports and dashed out early and gathered loads of "mums" that were in bloom before the early snow. I've been raking leaves and acorns. I hate oak trees for shade. I had bushels of large acorns. I hate to walk over them—like walking on marbles.

I feel relieved for I've mailed my Christmas cards and have all my presents except a doll for Alice. She can't seem to decide the kind. I do so enjoy buying a doll for a little girl. Bet I'll be buying her a doll when she is most grown, since she is my baby girl.

July 4, 1954
It has been more than a year since the Robin was here, and it was a year of much worry for us. We went to the brain specialist. He sent us back to the medical doctor. They decided it was a reaction from the serious operation he had for the cerebral hemorrhage. Ralph slowly was getting more nervous. They sent him back to the brain specialist this January. He decided it was scar

tissue from the operation and did an encephalogram on January 26, 1954. He drew spinal fluid and pumped air into the brain. This was to give an x-ray and shrink the scar tissue if that was the trouble. It was scar tissue, and it almost left him paralyzed. He spent seven and a half weeks in St. Mary's Hospital in St. Louis. I spent most of my time there too, as he was so sick. He has regained his health and has been very busy with the farm for the last three months. His case is a very special case. The doctors here just send him back to the surgeon. It seems like a miracle to have him so active and happy again.

After his cranial operation at age 41, Ralph was left with a pit in his forehead.

MJS: This session in the hospital was more serious than the original operation. When scar tissue would rest on the nerves in the brain, the body could stand only so much until it popped the tissue off and the pressure was released. The doctor compared it to something resting on a hot-wire. Hence, Dad would have a good day and then a bad day. He went back to St. Louis for help. That was when the doctor tried an encephalogram. This had not been attempted on a man who had the frontal lobe removed. It was too much for him, and he had to rebuild all of his responses. It took most of two months. This was a very frightening time, for one did not know what the future held.

The day this robin came, Ruth, our fifteen-year-old daughter, had an acute attack of appendicitis, and my mother was badly scratched by a mad cat. Alice barely escaped being attacked. Ruth had an appendectomy the next day, and mother started taking the fourteen "shot" treatment for rabies. Perhaps you can understand why I have not sent the robin on its way. Ruth is about over the operation, and mother has finished the treatments. They made her uncomfortable and nervous, so the twins have been helping her with her work. Alice has recently had her tonsils removed. We had Blue Cross and it has really paid, but I'm taking out Blue Shield too for all of this has happened in the last six months. We don't realize how happy or well off we are until serious sickness comes our way. For the first eighteen and a half years of our married life we had no serious illness in our family. Then trouble sort of caught up with us, but the outcome has been very successful. Enough of our trouble!

Our twins are home for the summer. It is a real pleasure to have them all here again. They surely have been helping me catch up on things I was behind on. We had an excellent gardening year. I made the garden while Ralph was in the hospital—he usually made the early garden. I surely miss that man when he gets sick! We have canned 165 quarts of green beans. We were "bean wacky"—it was the first we had raised in three years due to the drought! We have also been making slip covers for the living room; now we are starting the dining room. All of these things I just did not find time since Ralph was not well. I've had to help with farm problems instead. We surely need a son when such times come. A mechanized farm is too much for a woman to manage. My Dad always helps us, but it puts a load on him at his age.

I would like to send the letters on and hope they will return earlier this time. Perhaps I can sing a "happier song" by then, for really our worries have had happy solutions, and I made some real friends while waiting at the hospital.

June 10, 1955

It is raining now and very cool—too cold for the cotton. Crops are delayed and many are planting over. Aren't we farmers a complaining group? Ralph, Mary, and Alice have gone on a three-day trip to Urbana, Illinois to visit relatives, the first time Ralph has been away for over a year. He regularly has nervous exhaustion every other day, but does manage to accomplish a great deal by being very busy the day he feels good. This *trouble he has is mystifying to all our local doctors. The brain surgeon says it is scar tissue from the hemorrhage, and time will perhaps adjust it. In the meantime he feels perfectly wonderful one day, exhausted and very depressed the next, but he does keep farming.*

Martha is in college this summer, will graduate at end of summer with a B.S., majors in music and English. She will teach next year in Hazelwood High School in St. Louis County. Mary is home this summer, and oh so much help to me. Ruth is sewing, getting ready to go to Columbia to Band Camp. She plays flute and piccolo, is also pianist for our church. Last week she entered the "Sew-it-in Wool" contest and was an alternate to the state meet. Now she is planning to make a wool suit for 4-H club. How she loves a contest!

Blossom, how I wish you could come by if you come this way again. This polio vaccination seems rather confusing and disappointing. We had a case on the farm this winter—an eighteen-month baby left with one leg affected.

My cousin and family from Alaska are in the states this summer. We had a family reunion last Sunday and heard lots about the "land of snow." They are teachers. Their daughters are ten and eight years of age. How those children do want to come to U.S. to live. They were born in Alaska.

The Bermuda grass you mentioned in your household letter is a real pest. It just takes over. I have two small patches near the drive. I have killed two others by covering it with old linoleum all one summer. I've tried poison—didn't succeed. It will smother shrubs and flowers, and grows so vigorously you must mow every three days. It is dead with the first frost. Our blue grass yard is green all winter here.

November 1, 1955

Today mom and I attended the Extension Club Achievement Day. In the afternoon we had a lovely cotton style show. The pattern companies—Simplicity and McCall's—furnished the dresses that were modeled by our club women. It was lovely to see what can be made from cotton. Of course at this season *of the year everyone is busy picking cotton. Many mechanical pickers are used, but there is less waste with hand picking. We had four people who picked more than 300# today and the rate is $3.50 per hundred—just to give you an idea as to pay. Schools are dismissed, and many children pick and earn all their clothes. Ruth and I earn some by weighing. I enjoy getting out and dismissing the every day routine of housework.*

I have such lovely zinnias and marigolds blooming. I wonder why I worry about mums when marigolds can be so pretty and easy to grow.

Blossom, now I'll look at the "Progressive Farmer" with more interest. It is a good magazine for southern farmers. I accompanied Ruth to State Round Up. She made and wore a rose and brown wool suit. This is Ruth's last year of high school so guess she will ease up on 4-H. It was Alice's first year, so now I start the trail with her. I attended camp with her, as she was a bit young to go.

Yes, Ralph's exhaustion has a rhythm or pattern to it—regularly every other day. The brain surgeon explains it as scar tissue from the operation—tension builds up and throws the tissue off the nerves and he feels fine twenty-four hours—then twenty-four hours of nervous exhaustion. He has felt much better this summer, so I hope he is gaining. We have been to the best doctors in St. Louis, and they can offer nothing, only hope that time will help his nerves become adjusted. He looks fine the good day and by being very busy and efficient, he has kept the farm going. We don't have hogs now, but we do keep the cows. They are his pride and joy.

Ilene, how did the "store-teeth" work? I was happy to receive my new reading glasses yesterday. They make reading and writing a pleasure. (Wonder if they will help me get interested in all this bookkeeping Uncle Sam asks us to do!) Age is kind in some ways—my new vision has failed so I don't see the increase in gray hairs, moles, and extra hairs on my face until I took a look in bright sunlight with the new glasses. Now maybe I'd have been happier to have not had such a clear view of this face! Then too my white kitchen walls were not as white as I thought.

Martha is teaching music in a high school in St. Louis. She likes teaching, but misses home. Ruth will play in the band for the state Teacher's meeting in St. Louis this week and will spend the weekend with her. Mary is attending college at Jonesboro, Arkansas.

June 1956

You have been here a month! Do excuse me. It has been a month of usual activities in a farm home: garden, flowers, rhubarb and strawberries to put in the locker, house cleaning following a remodeling of the kitchen, graduation activities for our high school for Alma Ruth and for our teacher Jeanie. Jeanie finished her requirements at Cape College last summer and has taught as high school vocal teacher in St. Louis County this winter, but it was necessary for her to come back to Cape Girardeau for Commencement activities. Needless to say I've been busy. I drove to Jonesboro, Arkansas and brought Mary home from college, and we have had a busy week of sewing and getting her ready to return for summer school. Jeanie will come home today and spend a week; then she is off to Bloomington, Illinois to attend Wesleyan College and start work on her masters. Reckon they will ever get educated!

Ilene, you are the smart one to take organ lessons—made me wonder if I should take some piano lessons. Alice is our most musical daughter, but how she dislikes the practice. Jeanie has taken organ lessons this winter from Lee Miller studio, and they were $6 per lesson. She justifies herself because she had ten piano lessons after her school day to help pay the organ lessons she took.

Blossom. I have a completely white kitchen—covered the walls and ceiling with white tile board, so you might say I have a completely washable porcelain kitchen. After years of want, I have a maple table and chairs in the kitchen. Most meals are in the kitchen, and I did not enjoy the plastic deals at mealtime.
Alma Ruth will take Home Economics and journalism in college as a result of years in 4-H. She won the district achievement award in "Creative Writing" so that seemed to steer her toward

journalism. She did quite well in debate, so guess she has a natural gift of gab.

Ralph's health has improved; part of the time he has the nervous every-other day pattern, but it has lightened until he goes ahead much as usual. It has been five years this August since the brain operation. It seems nothing short of a miracle. The doctors agree. He looks so well and younger than I do. He says I get all the worry and waiting when he has made two trips to the hospital. Fortunately he doesn't remember the terrible headaches. Doctors told me to remember that intense suffering is usually blotted from the memory or "consciousness."

October 1956

You arrived with a busy harvest season. Combines for soybeans, sowing wheat, baling the last alfalfa, and gathering corn keeps the men busy. The school children, women and older men are picking cotton--three bales per day. It falls my lot to weigh at such time. At first I dread it--I blister from the sun so easily. After a few days I forget the clutter the house gets in and enjoy the open fields. What a lovely picture the fence row is! A mass of yellow golden rod with many black bees and orange butterflies to inspect them. As I never learned to pick cotton, I spend my spare time reading. What a treat! If I were at the house I'd see so much I needed to do. I'm enjoying being "field free" for a short time.

My girls have flown away again. September Mary Ellen went to Trumann, Arkansas to begin her first year teaching fifth grade. Martha Jean returned to Hazelwood High in St. Louis as vocal teacher. Alma Ruth entered Cape State College as a freshman. Only Alice and I are at home as Ralph was in the hospital.

Ralph had thirty-five stitches to close a wound just above his knee. A spike nail was removed from under his kneecap. He was caught in a power takeoff when they were baling hay. He spent fifteen days in the hospital, but walks without a limp now. He was lucky a man was near to rescue him. Farming can be very dangerous in this day of so many mechanical tools.

Martha Jeanie comes home often. She was always a "mama's girl" and brings Ruth on her way through Cape. How we enjoy those week-ends! Sunday night and Monday I feel so empty and useless. An empty house is such a lonely place. Some of you have faced that time too. What did you do (besides work) to adjust to this new routine? There is always plenty of work (for I miss their helping hands too), but I mean what inspiration to give life "zest." The farm keeps Ralph busy, and he doesn't seem at a loss. I guess I've lost a large share of my job, for the girls kept me busy. I do see I have more time for Alice; just the same I feel rather "useless" so often. Is that just a time in life or maybe "nerves"?

Jeanie's boy friend is home from the service (at present is working on his Master's Degree in Accounting at Missouri University) so it seems a wedding is in view for next June. I don't know much about such as it is the first, but I'm sure I'll get in on the sewing as her sisters are to be the bride's maids.

Ilene, I've had some bad times while Ralph was in the hospital at St. Louis, so I know what you mean, but I do like to go up for at least one opera per season. We saw "Kismet" this summer, also enjoyed Lowell Thomas' "Seven Wonders of the World" at Cinerama. Have you ever visited Shaw's Garden? It is the largest botanical garden in this country.

The day of mules and horses seems past here, so there is not so much demand for hay. We have such a long growing season here we can get lots of alfalfa from our acreage. Our saddle horse is getting old and doesn't have an appetite. She was reaching over the back fence to nibble the roses. She

Molly Brown comes to the house and expects chicken mash as a treat.

was near the laying house. I offered her some laying mash, and *she liked it. Now she waits for some!*

I am writing in Ralph's little office. I dared to paint the walls a soft yellow while he was in the hospital and lay a new rug. He has a nice collection of mounted insects, butterflies etc. on the walls. For fun I made "café curtains" of a printed material, gay with bugs and butterflies. Now the outside of the windows are alive with moths of all kinds crawling up and down and peering in at me tonight. I'd better close this before you decide I'm "buggy" too.

MJS: Ralph took eighteen hours of entomology in college and delighted in supervising all children and grandchildren in making bug collections.

April 1957
Ralph's brother is coming this afternoon, a family of five. We had planned to finish the bride's maid dresses while they were here to be fitted—now that is out with extra folks here. Mary and Ruth have orchid ones and Alice has pink—I'll be glad when they are finished! The wedding is June 16. The girls will be in school until the week before, so that will be a busy time.

Our Extension Club chartered buses and had a tour of Memphis last week. The azaleas were blooming—I didn't realize how tall they grow in the south. They aren't hardy here. We get them as potted plants. We also visited the factory and were shown the process of making Kleenex, Kotex, and Delsey tissue. The plant covered seventeen acres. I always wonder when I burn the wastebaskets if we Americans will always have such an abundance of paper. They assured us the tree plantings are exceeding the need. So go ahead burn the wastebaskets now!

Elora, we visited at Macomb, Illinois last week. I thought of you. Ralph has a cousin with cancer of the colon (had his stomach removed last year because of cancer), and he can't live. He is head of the Chemistry department at Macomb State and is still meeting his classes. He was so delighted to see Ralph. It had been eighteen years since they were here, and we saw them last.

December 1957

    My girls were all here for Thanksgiving, also the new son-in-law. Jeanie and Bill live at Poplar Bluff, Missouri. She has senior high school vocal music in a school of about 900. Bill is an accountant in a tax consultant's office. Mary teaches third grade at Lepanto, Arkansas and doesn't get home too often. Ruth is a sophomore at Cape State.

    Ruth brought four girls from her dormitory home this weekend so we had half a dozen girls counting Alice. Ruth has decided to major in Home Economics. Ruth and I were on the "Country Journal" program over KMOX this September. It was a leadership recognition program. I have led a club sixteen years. Nine of my "Yard Improvement" girls met this afternoon, and we made doorway decorations for our homes. We used limbs from a large evergreen plus red ribbon bows.

    At Thanksgiving we had already had 93 inches of rainfall. I suppose it will be more than 100 inches before the year is over! Our average is 44-49, which we thought was plenty in this level land! It is the worst year for farming we have had—cotton was too late so has not opened, beans were planted three times and are not all harvested yet, worms ate the corn up, pastures and wheat drowned out. It was too wet for gardens! It was just a year of failures for farmers here. It just rains and rains. The likes I've never seen!

    I would not allow the older girls to be majorettes or cheer leaders, but "in my old age" I have given in to Alice, and she is a junior high cheerleader. Has she ever practiced! It has trimmed two inches from the middle of her. Maybe I should try it. I like what the exercise has done for her figure, but I'm not too happy about attending all the ball games!

    Jeanie's wedding was lovely. We made the dresses, but had the florist decorate the church. I can't remember ever being so tired as when it was over! Then people came by the house for a week to see the gifts. They have a furnished apartment, trying to find a "reasonably priced" house. I have helped them refinish a lovely maple bedroom suite that they bought at a good price. We refinished it a" nutmegy" brown color. Remember my hobby

Martha Jean married Bill Stewart in the Lilbourn Methodist Church.

is refinishing furniture.

    I'm driving ten miles each morning to bring out a carpenter to put aluminum siding on the outside of the house. I cook his lunch and take him back at 5:00. Carpenters are scarce here, and that seems the only way to get it done. It rains so often he can't seem to finish. After living in a yellow house twenty-five years, I'll soon have a white one. Aluminum siding is expensive, but makes an old house new looking and eliminates painting.

July 17, 1958

    Everyone comes to the back door here, and then trails in through the utility room, so I have to keep the path presentable. I had a nice front entrance built when we remodeled. No one ever uses it except the girls. My son-in-law comes the back way too, so when they all marry, I'll lock the front door. Seven years ago we enclosed the back porch with casement windows and added pine paneling and a hardwood floor.

    It is time for the 4-H Fair. Alice has a dress to model, also a demonstration "A Practical Way to Cut Up a Chicken for Freezing." I'm glad she has it, for it is the best method ever. You come out with four meaty compact pieces that can be divided again to make eight, and a pile of bones for boiling for salad or soup. It is a new method from our State University. Ruth stages and narrates the style show. She is a junior in college in Home Economics, so this is a good experience for her.

I'm busy making pickles just now. Try this: wash cucumbers, pack in sterilized jars, add vinegar (1 quart vinegar, 1 cup sugar, 1/2 cup salt), seal. Ready to use in a few days. Easy and best ever if you like olives and their flavor.

I have high blood pressure so am on a diet to lose weight. It is a high protein diet so is mainly hard-boiled eggs. I feel like cackling I've had so many. I've lost 15# and feel lots better.

MJS: July 2, 1959, Joseph M. Heath died with leukemia. There were no letters in 1959, 1960 or 1961.

March 4, 1962

It seems January and February and staying inside just about gets my "get up and go" — so I visited the doctor this week. I look well, but how I ache and how nervous I get! Enuf!

It was a busy year. Ruth's wedding was lovely, and she is far happier than in college. She drives twenty-two miles and teaches high school home economics. They are both very busy, but most people are since so many women work outside the home. They eat Sunday dinners with us, and we enjoy both of them. David is a very inspirational speaker, the best our church has ever had. (I could be prejudiced, but others say the same.) He is such a pleasant person that you are happy to have him visit. Ruth helps with the music and helps with the youth work. All's well that ends well! I hope they stay here another year, but the bishop moves Methodist ministers about. So!

MJS: Ruth Hubbard married David Richardson who came to pastor the Lilbourn Methodist Church. He had just finished his seminary work at Garrett Theological Seminary in Chicago. This was his first appointment after graduation.

Ruth married David Richardson in the Lilbourn Methodist Church.

Jeanie and little Donnie were here today (their weekly visit). Jeanie expects a new baby in June. I'm hoping for a little sister.

MJS: Donald Ralph Stewart was born February 20, 1960, the first grandchild.

At Christmas I had my four girls and mother plus David's parents and Bill's parents. David's only sister is in Europe teaching while her husband is there studying on a scholarship. They had sent home lovely slides, so we spent the afternoon showing home movies and slides. It was a wonderful day, but the next day I was sick and I've "dragged" since. How did I get to be such a pessimist!

It has been a very cold winter for us. Ralph has planted a little garden and has fifty baby pullets. We can buy fryers dressed for 19 cents per pound, but we don't like the eggs we buy, so we keep a few hens.

MJS: I went to visit the farm and found mother harboring a box of chicks on the back porch. After she greeted me, she retorted, "Well, I don't know what your father has ordered, but it is some kind of wild chick." "How do you know it is a wild chick?" I asked. She responded, "They are little chicks, but they can stand still and jump eight inches straight up in the air." I could hardly wait to hear Dad's side of the story. When he came in, he responded, "There was an advertisement for pheasant chicks, so I ordered some. I thought we could put them in with the baby chicks." For several years we had pheasants on the farm until hunters decimated them.

Have you tried a blender? Jeanie has one from Sears. She uses it to make baby food. It is wonderful—makes the nicest slaw in about half a minute. I want one soon. I have a like-new, deep-fat fryer. We just don't care for deep-fat fried food, but I use it to blanch beans and asparagus for the lockers.

I am anxious to work in the yard, but it has been too wet and cold. I have lots of small mimosa trees that I need to give away soon. (We all share our extra plants.) We also have large cypress trees here—have given away several small ones. Most of you are too far north for these trees, I suppose. I have many shades of iris, but I am losing interest in them, so I'll try "mums" this year. Isn't it nice what a new interest can do for "lagging" spirits? My bittersweet just doesn't have berries; the nandinas have a modest amount. Jeanie's nandinas have dozens of bunches, large as grapes. I borrowed some from her for Christmas arrangements. Do you know what they are like? They are lovely—bright red. The crocuses are blooming now; first blooms come February 12 here.

October 1962
Several of you have had good letters in the Household this year. I can recognize your letters without looking for the signature. I thought I was observant of your style until I remembered this robin has been going for about twenty-five years. I should

recognize a familiar friend.

Blossom, your comments on the rain are true here too. The men work about three days each week and in the mud. We have used the machine cotton picker one day on the home farm. However, the people have hand picked several bales. I am glad there are machines to pick the cotton. It is such a tedious task. Our school system had a three week cotton picking vacation, probably the last year we will have one. We have had very few family vacations because, due to cotton vacations, we had summer school. Ralph was busy making or harvesting a crop or the girls were in school all year. School started in August this year. However, we got in an eight day vacation to Washington D.C., also saw Mammoth Cave, Jefferson Davis's and Lincoln's birthplace in Kentucky on the route going, and Annapolis and Gettysburg on the way back. I did enjoy it all, especially Mount Vernon (brought home a potted boxwood and bachelor button seeds from the greenhouse there). I'm keeping the boxwood in the house this winter. Congress was in session so we saw the Vice President in the senate. We enjoyed the mountain scenery going over, came home via the Pennsylvania Turnpike. It was faster but not so scenic. We talked so much about it that Ruth and David took the same trip on their vacation later.

Alice, I'm on your trail, or I have a granddaughter—the first one. She is Angela Ruth, has auburn hair and big brown eyes. She is a delight as she is so happy. She was born June 15. I stayed with Jeanie two weeks after she came home from the hospital—nice vacation for me.

I enjoy Sunday. I teach the junior class, attend church twice, and usually have guests for dinner so Monday I'm tired—but it is worth it. I believe they say it is "more fun to wear out than rust out."

This is my first year as Secretary of Missionary Education, so I needed to attend a School of Mission, but it came at our vacation time. I did attend the district meeting, and I have been reading "The Arm of Compassion" and "On Asia's Rim." Do we study the same books? David remained as our minister this year. As a mom I was happy to have Ruth and David near another year. She is teaching in our local school this year so does not drive so far. I can't comment on daughters-in-laws, but I can say we are very happy about our sons-in-law. Bill is so good to help with their babies, and he is an only child so it is all new to him. Present day youngsters want nice things, but it seems they work hard to get them. More women work outside the home to buy convenient household equipment. I did not teach after marriage, but my girls have both taught, and Jeanie gives piano lessons within her home with her babies!

We decided to raise fifty pullets for eggs, so we ordered Austra-white. They must have sent leghorns. They are small and can fly over the fence with ease! They roost in the tall oak and cypress trees! They kept the leaves pecked off two new roses until they died, but do they shine now by laying dozens of eggs. I only get 35 cents per dozen, so we eat lots of custard and puddings and keep the girls supplied. In spite of their digging, I have such nice marigolds and mums blooming. I believe marigolds are my favorite of the annuals—so many nice kinds.

At last I have an automatic Maytag washing machine! I have to have an iron remover filter so am short of white water if I use too much. I bought a machine with a suds saver and kept my old machine. I run the suds in the old machine and wash with both and rinse with the automatic. I like my old machine for Ralph's work clothes, slip covers, blankets etc. Mom says I need a coin laundry I wash so much. Two machines speed things up.

May 1, 1963

My thoughts have often been with Molly since the letter came. I thought of when we lost Mrs. Edna Eisenbise or "Hibiscus" from the old robin. I believe this group has been together for about twenty-five years now.

It is really spring here—asparagus to gather for the freezer each day. Today I set petunias and mums. It is that time of the year when days are not long enough to get it all done. Alice is graduating from high school on the 14th of this month, so there *is a busy schedule for her. The seniors here go on a week's trip to points of interest in the south—Vicksburg, New Orleans, Mobile, Florida—so that means extra sewing for her. For twenty-three years we have had a school girl in the local school. I fear I will lose interest in the local events; however, Ruth may be teaching here another year.*

I am to have the privilege of keeping my only grandson and granddaughter while my daughter and son-in-law have a vacation trip to Washington D.C. We enjoyed our trip so much that we urged them to go.

Donnie and Angel stay at the farm while their parents take a trip.

My mother is 73 but in good health, so we go to Garden Club, Extension Club, and WS of CS (Women's Society of Christian Service) together. She has an apartment built adjoining our house—just a utility room between us so we visit back and forth when weather is bad. She has a large living room, bedroom, kitchen, bath and sun porch. This way she enjoys complete independence, yet is in the family circle and we can see about her. We visit older people often. Several new retirement or nursing homes are being opened here. As people live longer there is a real need for such, especially when the elderly do not have a family. We visited a member of our church in such a home Sunday—wide doors for wheel chairs, hand rails in bathroom, nurse on duty twenty-four hours. All very lovely, also expensive, but it is fine to see the elderly so well cared for.

May 23, 1965

How nice to have you come again! It was two years since you were here last. It was like seeing old friends. According to my figures this robin has been going more than twenty-seven years. In 1937 we had a flood here. I wrote to the Household about our experience. Then Ilene invited me into this group. Much has happened in all our families during these years, and we have shared these times together, so I was so glad to have you come again.

Last week-end we visited an aunt near Springfield. She has reared seven children in Indiana and now moved back to Cabool, her girlhood home, to retire after her husband's death. She has published a book, really an autobiography. It was published by Vanity Press, so I fear she will lose money on the deal, but at least she is doing what she wants to do.

We enjoyed the scenery in the hills, but home looked good when we returned. About seven counties in Southeast Missouri are level without any hills. Any direction you drive, you see level fields of cotton, wheat, soybeans, some corn, and some pasture. We have cattle, so pastures and fences. Most farmers have large fields not fenced of row crops. Crops are up to an excellent stand but we do need rain. Gardens are poor. This is not too good a gardening area. It gets too warm, and our gumbo soil cracks and looses moisture.

Ralph bought a small horse at the sale yesterday to entertain the grandchildren. Jeanie has Donnie age five, Angela age three, and expects another in October. Ruth expects her first baby in December. They are so happy, for after four years of marriage, she was anxious for a baby. Alice will soon be home from college, so maybe she will enjoy riding. Ralph's saddle horse was a bit too fast for the girls or children.

The harder I work, the more optimistic I feel. How I dread those days of depression! So, I've been working like mad recently, but I am so slow compared to former years. I want to try something new or travel a bit to lift my spirits. Ralph likes the old and routine things. He dislikes any change. Maybe we balance each other that way.

October 1965

I helped Ruth and David move to a new church in St. Louis. She expects a baby in December. It will be their first, and they are very eager for it. They like the new church (Zion) just in the edge of the city. They have a nicer home than our parsonage—better heating for the new baby. Ruth has been home when David was away for pastor's school to have our dentist pull her wisdom teeth, then again as our church wished so to give her a stork shower. I am to go when they bring the baby home from the hospital to help those first days. Ruth said she would probably ruin my Christmas holidays this year. September 19 Jeanie and Bill had Teresa Jean, and I helped ten days there. It is such a delight to help the whole family get acquainted with a little new member. Teresa is growing nicely now. Then in August I helped Mary in organizing a four-room apartment. We bought some used pieces and refinished them and emptied our attic. I had so much extra bedding and extra of several other items. She enjoys a home instead of a room.

Last week Alice spent four days in the hospital and came out so weak I let her have the car. I stayed home, for the trucks are all busy in the harvest. This week I borrowed Jeanie's car to shop for a new car. So what? Today I was just too tired to go car shopping, and tomorrow is Garden Club. Maybe the next day. I occasionally have these tired days—not so often or long. It helps to go away from my routine here and to feel useful to the girls; maybe I get too busy to be bored and tired!

I am trying to Christmas shop early, so I can help with the new grandchild. I bought Angela a talking baby. They have such pretty voices. We play with it to hear it laugh. I can't seem to have ideas for the others so have bought books. I bought books they are always borrowing from home—"Fun Encyclopedia," "Leaves of gold," and "Best Loved Poems of American People."

Mom has raised a puppy, and it digs deep holes in the yard. I wonder why, as the ground is hard, and we have no moles. How can I discourage it?

May 30, 1967
I hope those who were ill are better, and how I do sympathize with Alice in her loss. Ralph's health has not been too good. He needs a hernia operation. I know it is lonely for Alice and Molly to live alone, but you can't run away from loneliness.

Mattie pampers Lucy despite the fact she digs holes in her yard.

The "expected baby" at the parsonage is a very appealing little lad of eighteen months. He is such a happy fellow. He was named for my father and his--Joe David. Ruth and Joe will visit soon while David attends annual conference of the Methodist Church. David's church is growing, and they are very happy in St. Louis.

Teresa will be two in September. She is a gay one and talks a merry clip! Her parents are having a new home built, so she visits us often to give her mother time for the many decisions of drapes and rugs etc.

Mary Ellen our twin age thirty-two suddenly

Mary married J.C. Jones in Jonesboro, Arkansas.

106

married. She worked in the college library at Jonesboro, Arkansas. I am not too happy for her marriage. The boy is twenty and has another year in the service. They are in South Carolina. I urged her to wait until she knew him better. They were engaged a short time, but of course, "Mother just does not understand." They have had trouble getting housing on the base, financial problems, and now Mary has health problems. Yes, and his mother has joined them, but she is a nice person, so maybe it will succeed. I've cried many a tear about Mary this winter. Now she is to have a cyst removed soon.

Yesterday Ralph and I took Mother back to the cemetery, ninety miles from us, to decorate her father's grave. We also visited the site of her childhood home and the farm, then on to visit a cousin. She had such a good time recalling her childhood. She has talked for hours. Tomorrow we will have a service at the cemetery where my father rests. These things touch us deeply and leave me depressed, but I must accept it as a part of life.

MJS: It was important to Mattie Heath to place flowers on her father's grave because he was the only parent she had known. We listened to the early stories of her life as she told us about the night her Pa had died sitting in a rocking chair beside the wood-burning stove because he could not sleep lying down. The pneumonia he developed after sleeping on the ground as a sixteen-year old drummer boy in the Civil War continued to plague him. Eventually he developed tuberculosis. Mattie was thirteen years old the night he died. She tells of how bereft and alone she felt. She had no home. Her older sisters were married. Her mother Jane had died when she was three months old after hanging up diapers on a wintry day and developing the flu. The mother's death-bed plea was, "Keep our four little girls together." The baby, Mattie, was sent to live with a cousin in Memphis who had no children. Following the mother's plea, Pa remarried to establish a home and sent for Mattie to come and live with her three sisters. She tells of her cousin's heart ache in giving up the baby they had learned to love and of their encouraging her to hide under the bed so she wouldn't have to leave the only home she had ever known. However, back to the sisters and Pa she went. The new mother did not last long, for she died in child birth with her first baby at age forty. Such was life in the frontier days on a farm. Mattie went to live with her oldest sister who married at eighteen. When the sisters had their babies, they welcomed Mattie's help, so she went from sister to sister. Memo was a story-teller par excellence. She delighted in words. I would often find her reading the dictionary "because words are so much fun." She loved to read the *Grimm's Fairy Tales* to us, her grandchildren, but we failed to understand how she could be so wrapped up in the gruesome tales. However, it was to her house I went with my college poetry book to get her insights into poems. She always had ideas that were born out of the experiences she had faced. What a delight!

Alice will soon be home for two weeks, then off to summer school. We will sew, wash and iron, and talk! How I enjoy having her home for a short time! She did two years as a music major, then changed to kindergarten work, so it will take another semester to do her practice teaching.

I have sewed today, made a blue linen dress. I don't enjoy sewing as I once did, but I get a better fit this way. I'm a little too tall and too heavy in the shoulder line for a size twenty dress. I like to cover my knees when I sit down. Present fashions are ugly to me—"bushed-up" hair and bare knees!

September 1968

Mary's mother-in-law died of a heart attack, and they came and stayed here and drove to Jonesboro and disposed of her household and house. Mary's trouble seems to shrink. We liked the lad she has married when they were here, so when he gets out of the Marines and grows-up a bit perhaps it will be okay. We are helping them. We think it is wise to do. Mary is 32, her husband 20! My husband is one year older than I, but he does not take a vacation very often. He seldom stays away over night. We did attend his niece's wedding in Chicago this month—away two nights, because I refused to drive 475 miles and arrive in time for a 3:00 p.m. wedding. I'm not a gad-about, but I would like to see a few places in America before this arthritis gets me too stiff to travel.

Ruth and Joe were here a month, but Joe had roseola and was very sick—had a temperature of 105 degrees for three days. I was to keep him while they had a vacation, but since he was sick they spent the time here.

Jeanie and Bill built the new house, so I had the little ones here often to help her. The new home is lovely, high on a hill that overlooks the golf course in the valley below. The lawn has been terraced and now Bill spends his spare time raking rocks and watering grass. He has even borrowed a stack of my flower magazines to read. He had scorned flowers before.

Alice will graduate in January. She is getting nostalgic about leaving college. Her boy friend has another semester. She will teach, for they need her even if it is mid-term.

The grandchildren enjoyed the neighbor's turkey this summer. Both little ones called them "gobbies," and both called the riding horse a "hor." Teresa was here last week; her great job and joy was to poke corn through the fence to the "hor."

I am more than a little discouraged by the war and the riots. With taxes $3 per acre we are farming for fun, but I'm glad I'm in America! We farm 1250 acres and have so little help.

MJS: The acreage of 1250 is accounted for with the home place of 945 acres, the Heath farm of 150 acres, and a farm by Risco, (Shady Grove) of 155 acres.

I've made a gingham dress for Angela today. I smocked it. I had almost forgotten how to smock. I priced a dress at a children's shop, and a very ordinary dress was $9.50! I decided it was time to sew again.

Tonight it is cold, almost a hint of frost in the air. I added the electric blankets to the beds today. I have poor circulation, so how I do enjoy my electric blanket! Ralph insists on a cold bedroom with an open window, and I shivered half the night until I tried an electric blanket. Now I give all the newly-wed nieces and nephews electric blankets for wedding gifts.

January 16, 1969
I've just "plowed through" a box of Christmas cards, a few bills, plus a few newspapers that I had stacked during the Christmas-wedding season (a big box of 'em)! Now, I know I bragged that Alice did all the wedding planning. She did, but that left the "worrying part" to me, and worrying can be time consuming too. What did I worry most about? Yes, I know you can't change the weather, but you can worry about it. Well, the weather cooperated for the wedding night, but we surely were afloat the night of the rehearsal. Now that is over, I'm open for suggestion for new thoughts or worries. Alice and Dan came last week-end and opened thirty-two gifts and cleared the sun porch of boxes! That ends our weddings! (No repeats, please.)

MJS: Alice Hubbard was married to Dan Bourzikas, December 28, 1968, in the Lilbourn Methodist church. Both were teachers in St. Louis.

[Date Obscured] 1969
Ralph with sixteen stitches for a hernia repair and my mom with four stitches in the back of her head, plus a concussion kept

things a bit on the serious side in this household. Both are better after a week for Ralph at Sikeston Hospital, and mom a week at Cape Girardeau Hospital. Ralph is quite active, that is, he is on the "Sale ring" again since it rains twice a week, and he is not encouraged to farm. Mom has had to be very quiet or she was dizzy, and she is so afraid of falling again. For two weeks I stayed at her apartment all the time. She is up and about some. Then arthritis showed up in my feet and slowed me down a bit — so I visited a doctor. Now we are all going great guns.

MJS: The "sale ring" was the farmer's mode of entertainment during the winter. All the implement dealers, seed stores, research specialists, and marketing gurus hosted dinners to tout their products. Farmers entertained themselves by taking in all the freebies.

I've helped welcome home to the parsonage one new baby, Stephen Wayne. He was a good one and did not cause much trouble. I was "Mama Bear" while Joe was "Papa Bear." Then, I was "Wilma" while he was "Barney"! What an imagination!

Mary does not seem to want us to come. J.C. (Mary's husband) is to come to U.S. in May. Poor Mary is so confused, but it is something for them to settle. We have offered help. That is all we can do. We would love to see Mary and Mark and hope they come home. I can't see any chance for Mary or Mark to be happy with J.C.

April 2, 1970
It was cloudy and cold outside today, but it was warm and a happy day at our house. We attended Sunday school and church as usual, and enjoyed seeing Mark (and Mary) lead the Easter parade of little ones with their "mite boxes" to the altar. Mark enjoys going to church. He looked so nice in white shirt, bow tie and plaid coat. Mary took care of the little ones in Sunday school today. We have enjoyed Mary's presence so much this winter. She had a successful operation and is much healthier in many ways now. I took care of Mark for six weeks, but she cares

for him now. He is a good fellow and very easy to manage. He really brightens our home. He will be two years old May 13. He is beginning to talk.

MJS: Mary came home with baby Mark and scheduled a much needed hernia repair. Her marriage to J.C. was in shambles.

Jeanie, Bill and the children came about four p.m. and stayed until eight p.m. Jeanie will visit Ruth this week and bring her to visit us while David attends a special conference. Ruth is feeling better. We have worried, for the doctors could not find her trouble. Stephen is a lot of work, crawling, teething etc. Three grandsons really made this a gay place—a four-year old, a twenty month, and a year old! I miss them when they go.

My Extension club of eight members met with me this Wednesday. At one time there were twenty members. I've been in this club twenty-four years. We are growing old together, but we do enjoy the fellowship. We had a lesson on the new fabrics and their characteristics. Cotton has lost the market to the man-made fibers.

We have had far more snow than usual this winter. We have a new furnace—central heat and air—we have enjoyed. This, plus wall-to-wall carpeting made us very cozy. We needed it with the little ones here. It is the first winter I've really been warm, and it is the coldest.

March 17, 1971

Mary continues to work nights, so my day and nights consist of time spent with Mark. That suits me, for I find I care less about outside activities (lots of

Mary and Mark find living on the farm with grandparents comforting for all generations.

them are "purposeless," tiring, and sometimes boring). Mark is the most teasing child I've ever known. He talks and talks, plays ball, record player, and boxes "at" you when you read. He makes much noise, laughs uproariously, cries in like manner, turns summersaults, cartwheels, and keeps things cluttered with toys. Time is never dull here. In loving Mark one evening I called him my "sugar plum." The next morning he responded by lovingly calling me his "sugar bowl." So! Ruth and her boys may come with Alice and Dan next week-end. Ruth will stay a few days. She warns me it will be hectic with three boys!

Anyone in favor of starting a new political party? I'll lend you my support. I'm so disgusted with Nixon and his expanded war.

February 1, 1972

Someone asked how old was the "Robin." My twins—Mary and Jean—were past three years of age when I wrote. They will be 38 years of age this February 12. So much has happened since the letter was here.

What here! Our baby Alice has graduated from college, married, lives in St. Louis, and both are school teachers. We have grandsons—Joseph David, Stephen Wayne, and Mark Edward, soon to be four. Mary and Mark came home when Mark was eighteen months for Mary to have surgery while I kept Mark. Her husband wanted a divorce, so after two years, she agreed. She took a nursing short course and worked as a nurse. Mark stayed with me of course.

Then as a climax to five years of being tired, I had a heart attack. I spent four days in the intensive care unit followed by a month in the hospital. That month was not too bad, but I came home, developed a bad pain in my back, the next day returned to the hospital by ambulance to have the doctors suspect a blood clot. After a special lung scan, the report was that the blood clot covered 60 to 70% of my right lung. I began to cough blood. At 10:30 p.m. on July 6th, I had surgery. Clips were put on the vena cava to catch the clots. And the veins in my legs in the groin area were tied. I had a slow time recovering from this but at least my heart survived this. I wear strong elastic leotards from my toes to my waist, sleep with my bed foot elevated five inches, and have a nice thirty-two stitches down my tummy! The doctors say that after a year, I should be all right. I feel better each day. Our girls all came home, and one stayed at the hospital around the clock until I was better and they called a private nurse. Mary gave up her job, and she took care of Mark and me. She is happy to be here with Mark. My mother lives in an apartment adjoining our house, and her vision is failing. So you see we all need help. We enjoy Mark so much. He is a very blond, happy little boy and adds sparkle to our days.

Ralph had a good crop year of cotton and soybeans. He takes care of the Hereford herd and has just been in to report nine baby calves. Ralph had brain surgery twenty-two years ago, and we wondered if he might need to retire early. However, he has good health except for some nervous, tired days. He is sixty-two and I am sixty-one. We have a record—he has silver clips on an artery in his head, and I have a dacron clip in the veins to my heart. I do not fear death as I did, but I surely do appreciate living. We live and learn, and little dream what the next years hold for us. Facing death gives one a greater appreciation of all living.

Ralph brought Mark some new boots. He dashes out to show them to me ever so often.

I have just finished a big seed order to Atlee Burpee Company. Seeds are high, but groceries are higher. No, I'll not be able to garden, but I can cook.

MJS: This final letter from Alma comes from a note that she wrote in Ralph's Round Robin.

February 6, 1972
I've just sent my other robin on its way. I've written in a farm robin with women from six states for 35 years! It was recently lost for four years. The "writer" from South Dakota lost it in the back of her desk. We had many mutual interests. Our children were small when it started, and now some couples are retiring.

Speaking of retiring, perhaps describes what I've been doing since June 7. I've really enjoyed this January. I've read, read, read. I've stayed home to avoid the flu. Recently, doctors asked me to take care for a year. Mary leaves Mark with me to avoid chicken pox, etc. We have Sunday school here, and Mark assures me he really enjoys our Sunday school! He is enthusiastic, loud, complimentary, and sometimes stubborn. I guess he is an average boy. His clatter keeps us amused. Maybe his ideas have helped me enjoy this January. (I usually find January a dreary month.) Then, too, illness gives you a greater appreciation for just simple living again.

We had a great Christmas season. All were here at various times except David. His mother had a stroke last year and is in a wheel chair, and his dad broke his hip so he was needed there. They live in Kansas City, but David brought them to St. Louis. She is in a nursing home. Mr. Richardson has been with Ruth and David after he came out of the hospital; he is very anxious to go home.

Jeanie's children are having the flu, one at a time--so we miss her weekly visits. Saturday will be the twin's birthday. We hope all are well, and we can go for a birthday party.

My mother's vision is failing, but otherwise her health is good. She has "talking books" and a player. They are truly wonderful, for she enjoyed reading so much. There is a great selection of books and magazines available. If you know some elderly person who can't read, tell them about the service. It is free for blind or partially blind.

Alma Hubbard

Chapter 6.
Until the Last Kick-Off, Ralph's Letters
(1964-1982)

Ralph C. Hubbard wrote letters in a Round Robin to his four brothers in Illinois:

Martha and Fred Hubbard pose with their five boys, including Ralph behind his father.

(1) Robert K. Hubbard of Urbana, married to Faye with five children, Julia, Elizabeth, Margaret, Dorothy, and Nancy. (2) Linus O. Hubbard of Chicago, married to Geraldine with children Sylvia and Robert. (3) David F. Hubbard of Urbana, married to Frances with children Fred, John, Kay and Carol. (4) Joseph E. Hubbard of Peoria, married to Donna (divorced) and Irene with children Joseph and Jenny.

Martha Hubbard looks fondly at her boys from youngest to eldest – Joe, David, Ralph, Linus, and Bob.

MJS: Ralph's letters bridge the gap between Alma's letters and give us a picture of the mundane farming activities from 1964 to 1982 and the camaraderie he shared with his brothers. The letters have been excerpted to reflect an emphasis on Ralph's farming operation.

May 24, 1964
Lo Folks,

Well I see some men now write in this robin. A month or so ago when it arrived I assured Alma that the responsibility for a letter was hers, and she said it was mine, so here it is.

We too have had unusual happenings this spring. First, we had the driest winter since 1930, but then big rains for a week and the farm became a lake with the highest water since 1945 when the farm also was a lake. Both this year and then, water poured over U.S. 62, and we used my rowboat to take care of cattle and calves at five of the six barns on the place. It was fortunate that thirty-six of the calves had arrived during January and February, for we did not lose a one due to the flood.

The past two weeks Alma has had a carpenter working on the house as she did last spring. This spring the front rooms in the house and the kitchen have had a face lifting. Last year it was our back bedroom, the bathroom, and the utility room. Since these events take place while I am planting a crop, all is confusion for a while, but now things are almost normal again. The crop is planted and up. Most of the cotton has been plowed twice and hoed. We are cultivating corn and beans with stands a little better than a year ago, although there is still water in many ponds.

Joe, I enjoyed seeing the winter picture of your farm land and the one of your sheep.

Yesterday Fred [David's son] arrived from Ft. Leonard Wood, and we enjoyed his visit here.

I notice the Illini [University of Illinois] have now lost thirteen baseball games in a row. Although a year ago they were conference champs, so football is not the only sport in which they may be very good or very bad. Well, the good book says, "I would that ye were either hot or cold," so I'm sure Illini athletes are very religious.

P.S. Alice will be home Thursday to spend the summer vacation with us.

November 30, 1964
Dear Chirpers (Robin),

I have just read your letters and was much pleased to hear, Linus [that] you are better, Bob of your trip west and the new grandson, Dave of your vacation in Wisconsin, and of Irene's nice letter with Joe's note.

The weather here is dry, and work is about on schedule again. Yesterday I had two colored crews of cotton pickers and our own crew of white pickers, about seventy individuals but working in different places on the farm. We were behind with the cotton harvest due to excessive September rains which had resulted in much mud, water, and large mosquitoes. We also were operating two large combines on beans and had one man gathering corn at the same time.

Jeanie, Angel and Donnie are here this week while Bill is getting ready for the C.P.A. exam. Donnie and I made a Jack-o-

lantern out of a pumpkin last night. . . .

Our big pecan tree in the front yard is loaded this year, and the crows have been having a nice breakfast. . . .

When I think of all the travel and hot air that has been wasted trying to get votes, I am happy that I am not a politician. There must be some awfully disappointed candidates over the country today. Before I came here I heard they had Republicans in the Court House, but for more than thirty-four years now there has never been one elected in this county. As usual I didn't lose all my votes, so I'm still happy and hope all of you are too.

MJS: New Madrid County was controlled by a political machine. All candidates were elected by the Democratic machine. I was told that selected voters were given a bottle of liquor and trucked from poll to poll over the county. Registration was non-existent.

February 26, 1965

Two days ago it was just like spring here, but this morning it is close to 0 [degrees] with snow on the ground. There is even a drift back of the house 2 ½ feet high.

Mary has started housekeeping in one-half of a double house, and Alma and I paid her a visit Tuesday; we took a few things from our attic which she thought she could use.

We have 28 new calves now, two of them in January and the rest in February, and as yet we have lost none this year or last, so I still have a chance for another 100% calf crop as in 1964. Joe, we were interested in your dozen lambs.

MJS: Four of the five Hubbard boys had agriculture degrees, so they shared a common interest in farming.

Well spring is just around the corner, and I am ready for it. Usually I make a garden about Washington's birthday, but this time I seem to lack the urge. I see a two week time limit suggested on the Robin so guess this should be on its way.

June 21, 1965

My goodness how this Robin flies around! It comes back around in less than a month now where it used to take three months; that shows we must no longer be folks who put off things, and we like to hear from the rest of you. Which letter in the bunch do you read first when it arrives? This time old self-centered me reads my own and was surprised that it was written on the 23rd of May, less than a month ago.

Yes, we have a new minister since (our son-in-law) David has left for St. Louis. His text Sunday was the First Psalm, and he preaches loudly like the good old Baptists I used to hear when I was a boy. I remember one time when I was a child, the preacher called on Dad to recite the First Psalm, and I was surprised that he knew it, so it became my favorite passage. I would feel honored if they used it at my funeral which I hope is still many years away, for this is still a mighty good world to live in.

August 3, 1965

Alma went home with Mary Sunday to spend a week in Jonesboro. Sometime next month she will spend time in Poplar Bluff when Martha's third child arrives. At present, of course, Alice is my cook and housekeeper. Last week she was song leader and counselor at the Methodist Camp at Arcadia.

The farm is very dry, and a rain would be most welcome. We cut and bale lespedeza hay nearly every day, and also use hooks to weed the beans which are four or five feet tall. Corn is nearly ready to gather, and I expect to start feeding some to my steers. Cotton should be ready to pick in another month.

Sunday Jeanie, Bill, Donnie, and Angel were here for dinner, and the children enjoyed riding the pony and going with me to run pumps at the center of the farm.

October 8, 1965

A month ago I thought I was going to finally get ahead of schedule and seeded forty acres of wheat, which I thought might be a foot high by now, so that I could have a fine early fall pasture. A rain drove us out of the field and turned out to be seven inches in thirty hours, which, of course, turned the farm into a lake, and

I wore hip boots and rode the little horse over the farm while the water went down. One of those mornings I saw two big bunches of wild ducks and geese, and I wondered if it were going to be like old times when they stayed here all winter.

I see, Joe, that you and Linus have had patch-up jobs on your anatomy, and I am glad to know you are both better. Health is priceless and essential to happiness. While at times I too have aches and pains, I'm able to stir about on the farm between certain TV programs which I must observe to properly rest and pamper myself. I can report that I too am really in good condition and no longer ever underweight.

Sometimes I think the things I do really do not amount to much, for I may patch up a barn or fence, recondition a combine (for in a few days the beans are scheduled to come out), or see how the cotton pickers are getting along. I seem to spend most of my time here, for as I have told others, I have not looked for a job since I arrived here. You know in 1931, I saw certain school board members about a teaching job near Oakland and at Lovington, Illinois, but it seemed neither place thought my future should be with them, so here I am.

My man who feeds the cattle and plants (seeds) has been at a welding school in Malden for a month now and will continue there most of this winter. I cannot be sure that he will be here next spring when this job is completed.

We had a 100% calf crop in 1946 and 95% this year which is probably better than most cattlemen do. I do not think I could enjoy farming without the cattle. Last Thursday I watched them judge the Herefords at the Mid South Fair in Memphis. They have a new building just east of the hog barn similar to the Illinois Assembly Hall which will seat 12,000 at a basketball game, also a new stadium nearby which seats 55,000, so you see development is here too. It took only two hours to drive to Memphis.

In the last five weeks we have more rain, and the place is like a beehive of mosquitoes. We have big ones, small ones, black ones, brown ones, and they all are very active and hungry. Last Sunday our pastor asked us to read the Book of Job. As these pests gathered on my ankles, neck, face and into my eyes, I thought of Job, but, why oh why, should I complain!

We have a new grand daughter in Poplar Bluff. Ruth and David visited us the past two days. As you know Ruth expects one in December. My new granddaughter looks just like her big brother Donnie.

Yesterday I listened and watched the second game of the World Series on T.V., and how I enjoyed seeing the Twins whip the Dodgers. I also enjoyed reading of Urbana's great football team, and how I will gloat if they beat Champaign.

December 5, 1965
...I was interested in hearing about Joe's Christmas tree business at Peoria. Several pine and cedar trees have volunteered here in different fence rows on the farm. I guess the seed floated here or was carried by the birds. We also have some squirrels near the house and have enjoyed watching them get a share of the two pecan trees in our yard. I mention them because until the past year or two there have been no squirrels on the farm since 1931.

January 17, 1966
...Today I expect to attend the annual Soils and Crops Conference at New Madrid, and tonight a supper of Methodist Men at Lilbourn. Yesterday I watched the East All Star Pros rout the West in Los Angeles even though former Illini star Bletkus played for the West. The color was beautiful on T.V.

Since I last wrote, I have been twice in St. Louis, once a few weeks before little Joe arrived and once since. I held the little guy when he was five days old and marveled how one that small could look so much like his father. Ruth now reports that he weighs more than nine pounds, so he is two pounds larger than when I saw him. Jeanie brought over her three children Saturday. The baby (Teresa) now weighs sixteen pounds, smiles and enjoys turning

"Our baby can beat up your baby," says Donnie Stewart, looking at new-born Joe Richardson and sister Teresa, three months old.

123

her head from side to side when you hold her. Donnie and I kicked a football around in the front yard when we were not watching football on T.V.

Linus, I was interested in your report of the "Youth's Companion." I never was the reader that you and Bob were, but can remember its appearance at the house regularly. This winter I have been reading "Fifty Years Below Zero" by Charlie Brower. This book was sent to us about eight years ago by Alma's cousin, Sterling Crowell, who has been teaching in Alaska for more than thirty-two years.

Bob, now that I have two grandsons the same as you, perhaps I should report on their activities. Donald Ralph has been riding a bike for some time now, and Joe David smiled at me when he was five days old. Now my oldest granddaughter, Angela Ruth, tried to tease me Saturday afternoon as I watched Buffalo lose to the All Stars on T.V. She is as full of life as can be and switched programs or turned off the T.V. for my benefit.

Dave, I am glad to hear you and Bob pitch horse shoes from time to time. I believe if I practice enough I might still throw around 25% ringers, for I threw one out of four yesterday morning.

May 6, 1966

Yep, it is planting time here. Perhaps this may be the latest crop since 1945, the year Alice was born, for as yet we have it all to do because April was extremely wet. We can start planting cotton on the Risco place [Shady Grove] today, for it is sandy there.

Ruth and little Joe David spent a week here two weeks ago. He is nearly three times as large as when he arrived four months ago and in every way looks and smiles like his dad even as Donald Ralph, our other grandson, looks just like his father too.

I see by the new football schedule that Illini will open the season with old Mizzou in Champaign. I notice too among the football notes about spring practice that Ralph Waldron might be a defensive halfback on the Illini this fall. His dad and I used to enjoy wrestling in the silo when we were kids. Usually Ralph

Waldron Sr. would get one of my arms up behind my back, and I would have to give up. His boy is six-foot two, and one of the best all around athletes ever produced at U.H.S. [Urbana High School]

Alice came home yesterday and will be our guest this weekend. Tomorrow we go to Poplar Bluff to see Martha's baby christened. She is still as blue eyed as Martha. Yes, since last I wrote in the Robin, we have been to Urbana to see Illini and Iowa play basketball, and also made a trip to St. Louis to bring Ruth and Joe here, so we two get around some.

May 13 1966

I find it has now been one week since I wrote, and we now have the cotton and corn planted and about seventy acres of beans.

Yesterday we were at Poplar Bluff. Donnie has had the mumps and chicken pox. Teresa is so blue eyed and loves to be held. She can now sit alone and reach over and pick up things and has two teeth. Angela Ruth is surely one bundle of energy and go. She is quite a talker too with a voice that carries distinctly, perhaps like my dad or Mrs. Heath. Alma thinks Teresa will be small like Mrs. Stewart, Bill's mother.

Joe, do weeds grow as profusely about your place as they do in my garden this year? The past month the lawn needs mowing at least once a week, and since it has been so wet south of the drive, I think I just as well harvest hay there in June.

October 24, 1966

...Ruth and Joe David are here with us this week. He stands alone now, but only takes a few steps, but how fast he can crawl! I gave him a little football about twice the size of a tennis ball which he throws with either hand and then crawls quickly after it as though to recover a fumble and then throws it again. Last night he enjoyed the little rocking horse his mother rode twenty-five years ago.

We may start bean harvest in a week or ten days. As yet we have not had a freeze to kill tomatoes, and the beans and cotton *are still growing. Corn has been ripe for some time. It has been wet here most of the time since August, and mosquitoes get thick at times. We have managed to plant wheat, but, of course, none of it is behind beans which I like best to do.*

Joe, I notice you mention your ponies eating tomatoes. Do you have more than one? I have a Welsh pony which I ride when it's muddy. It is only half the size of the horse I used to ride, but sometimes I think she is faster. When she opens up, my pickup truck at 70 seems slow.

April 14, 1967
... Again Ruth is here with Joe David who now insists on unplugging all electrical cords in wall plugs. I'm afraid he will electrocute himself. He is now quite a prankster. The other night I lay down on the floor to play with him when he comes up and bangs me over the head with the wastebasket.

Bob, I share your enthusiasm, for when the kids are here with the grandchildren and then they leave, it seems so quiet. The joys of being a grandfather are great.

Alice, Dan (her boy friend), and I caught insects last Sunday, for Alice must collect them for a course she has this semester. Dan is from St. Louis and a music major. He plays the piano most vigorously. It will take both he and Alice another year to get a degree and perhaps both may try to teach in college which might mean endless education.

Pasture conditions have been so good I lost a cow from clover bloat Monday. It was the first bloated in ten years.

I pitched a few horse shoes with J.C. when Mary and he were here, but do well to hit one ringer out of ten now. Too old, I guess. At least I can still beat my sons-in-law though.

June 19, 1968
It seems that it has been more than a year since I wrote in this robin, and I feel ten years older. In three more years, I will have farmed here forty years, and I think it will be time to quit.... Joe, have you played the flute or piccolo since you went to Peoria?

We too had quite a bit of wind here the night thirty-four

people were killed in Jonesboro, Arkansas. That wind split our pecan tree in the garden east of the asparagus bed, and a week later another wind broke the other half off, so only a stump remains. That same wind took the roof and half of the chimney off the old house at the north center of the farm and scattered tin and bricks over forty acres. Yesterday when Johnny was combining wheat one-fourth mile southeast of the building, I picked up three more pieces of tin from that building.

Two weeks ago, Sunday evening, we saw Alice graduate at Cape. They had the ceremony at Houck Stadium, and when I arrived the graduates were in two groups at each end of the football field, about 375 in each group. It was quite colorful to see the groups come together and sit in chairs for the speeches. Alma tried to make a movie of the event.

Since June 4th we have had a big light, south of the house on a high pole, which makes it as light as the old street light at 801 East Oregon [Urbana] when I lived there. At first I thought I could not get used to the moonlight every night but guess it is more like old times or what others are used to in the city. Now I guess I could work in the garden at night.

Since it would require a 2800 mile trip, I have no idea when we will see Mary's son. She says he is blond, long and grows longer, so perhaps some day he too may be six feet three like his dad.

October 11, 1968

As yet we have had no frost, but during the past week each morning, I could see my breath. One year on the sixth of October, it went down to 18 degrees here, so it could freeze hard anytime.

Mary and Mark are still at Key West and okay we hope. Mary started teaching twenty-nine first graders. J.C. is still in Vietnam, so he is half way around the world.

Joe, we too baled hay even a week ago, for rain has been often here every month except July which was very dry. I combined some Custer soybeans two weeks ago and some Hill soybeans last Saturday. The Picketts and Ogdens will probably not be ready for a month. Custer is an early bean and like Picketts resistant to the Soybean Cyst Nematode. Custer pops out very easily though. . . .

We enjoyed so much the visit when Linus, Geraldine and Robert were here. I can always see things which need to be done, but I can always put things off now. Is that a sign of old age or lack of enthusiasm or just plain laziness?

MJS: Linus was the only one of the Hubbard boys who did not major in agriculture. He was an electrical engineer and repaired all the mechanical things that had problems when he came to visit.

Alice is to be married next spring, April 4, to a boy she has known at Cape for four years. At present both she and he are teaching in St. Louis. We would like very much to see Mary's son and may go there sometime before Christmas.

January 16, 1969
. . .I do have five small calves; the first one arrived January 9. I have also attended a few farm sales which are very numerous this time of the year.

We enjoyed very much Alice's wedding and those whom it brought here. Last Sunday Alice and Dan were back, and Jeanie, Bill, Donnie, Teresa, and Angela were here. We really filled a pew in our church Sunday morning. How we always enjoy company!

April 21, 1969
It's been three months since I've written in this Robin and much has happened. Right now is the time to plant crops. A month ago today I was repaired at Sikeston, and that week we had plowed four days, but it has not been dry enough to work in the fields since.

MJS: Dad had a hernia repair, an affliction inherited by most of the Hubbards.

Yesterday Jeanie's family came over from Poplar Bluff, and since the sun shone brightly, we were out doors most of the afternoon. Angel and Donnie fished some (in the stock tanks) and rode bicycles. Angela rides the same one her ma rode here thirty years ago. Bill and Jeanie dug up flowers and good dirt to replant them in their yard at Poplar Bluff.

Mary writes that she is packing up to go somewhere when J.C. comes home in May. We hope it is here.

Alice and Dan have been home twice since their marriage, and I have been there twice, so I am familiar with their place and can drive there easily when in St. Louis.

Ruth writes that she may bring Steve and Joe here for a time in May. It certainly makes it a happy place here when some of them come back.

I feel like I am in style since all of my brothers have had hernia operations. It seems odd that for fifty-six years of my life I escaped such a family weakness. I have really enjoyed letters from all of you and hope all of you are well.

The Mississippi is to crest the day after tomorrow at New Madrid at about thirty-five feet which is twelve feet lower than it was in 1937. *[The big flood]*

August 1, 1969

... Yesterday I took cattle to Charleston, and some steers brought 29.40 and six heifers at 26.90, but they were light for my pastures are short.... I am to meet Dan at the Methodist men's retreat at Arcadia and will bring him back with me. Perhaps I will again pitch horseshoes, for part of two days there. Bill and David will be at Arcadia too... While at Arcadia, Dan and I went on north fifteen miles to Caledonia where we saw a $100,000 Polled Hereford bull and some of the finest cows (Herefords) I have ever seen anywhere. Their pastures were wonderful compared to our burned up ones here, for they had been having rain. Last Sunday morning early I went fifteen miles southwest of Arcadia where the Union Electric Company has built a fifty-five acre lake on top of Taum Sauk Mountain, the highest point in Missouri. All *the water in this lake is pumped up by electricity and then makes electricity by merely pushing a button when they need it when necessary. It certainly was some way to store electricity.*

November 20, 1969

The Robin arrived here with an inch of snow after five inches of rain since Apollo 12 sailed for the moon last Friday. Usually in the wintertime when they blast off one of those things I receive a four inch rain. This time it was five inches. Last summer though when I really needed rain I only got an eighth of an inch, so I don't know yet whether they have discounted how to make it rain or not.

Today I would like to go to the Farm Machinery Show in Memphis. This is the last day of a three day show. I notice, Joe, in your letter you had a hard freeze on October 19, which was the same time we had one which killed tomatoes here.

The past month Alma has taken care of Stephen Wayne, Ruth's youngest child who will be nine months old November 25th. He doesn't walk yet but crawls rapidly and pulls up and pulls off things when he can reach them. He has eyes like my Dad and lots of vigor and so much twist and turn it is very hard to put any thing on him. Sometimes he is like an owl at night and likes to stay up or come up around one o'clock. His mouth seems to be open whenever his eyes are open. Ruth is better now, so in a few days she may come for him. It will then be Mary's turn to bring Mark Edward here who will be two in May. Mary wants us to keep him while she too has an operation for a hernia, the weakness of most Hubbards.

Well, I did go to Memphis to the Southern Farm Show and saw tractors from West Germany, England, and Japan in addition to all sorts of American made models. I never saw so many gadgets for the farmer to buy, most of them very impractical and at ten times the price they will ever be worth to a farmer, but it was interesting anyhow. The complete floor of the Coliseum, a building the size and shape of the Assembly Hall at Champaign, and huge exposition buildings were filled with displays.

Today we go to Memphis to meet Mary and Mark who are

to be here from North Carolina. Yesterday we took Stephen to Perryville where we returned him to his ma and pa. Perryville is 125 miles north of here so by meeting Ruth, David, and Joe David there, it made a good way to return their son.

December 2, 1969
Lo Folks,

Here it's the last month of the year, and we are still harvesting soybeans. It's nice to still have beans to harvest, but we lost several days when we had a five-inch rain during the time Apollo 12 sailed for the moon.

Mary and Mark have been here eight days, and Mary has an appointment with Dr. Heeb at Sikeston December 11, who operated on me last spring, for she too has a hernia. We are to take care of Mark.

Mark weighs 25 pounds and throws well with either hand although usually with his right one. He is as blond as Mary and I, with hair as white as snow. He has a better appetite than any of the other five grandchildren and might grow to be tall, for his dad is six foot three.

When we met Mary and Mark at the bus station in Memphis at 9:30 A.M. on Sunday, a blond man carried him down the bus steps before I even saw Mary, handed him to me and said, "Here is your grandson." I certainly was surprised, for he seemed large for only one and one-half years old. The little boy looked half amused at me as I carried him to the car but said nothing...

April 2, 1970
...The yard is full of robins; they stop and rest here a few days but seldom stay here long enough to build nests. I also saw a jay bird in the tree south of the house, and red birds have been around the past six weeks, so I know spring is here.

Ruth is here now, and we again enjoy the three sons at our house. Steve crawls out of his high chair and up on the table when he eats.

July 7, 1970

I have just finished all your letters and very happy to hear from you. Joe, I read where you made hay six weeks ago. As yet I've never pulled the hay baler out of the shed, for I just finally quit planting. It has, however, quit raining and I should make hay next week.

This year has been much like the year Alice was born when the farm was a lake the 16th of June. It rained nearly every day the first half of June and then again the night of July 3. Since then it has been dry and the ground is hard and cracks are appearing between rows.

I now have two teeth to fill the hole where I broke two out some weeks ago. They look well, I guess, and I wear them much as I would a tie to improve my appearance.

At present Ruth, David and their two sons are here so again, so interesting to watch the three boys play together for they have such a good time. I'm sure when I was Steve's age, Bob, Linus and I must have caused a lot of interest in our home then.

Ralph, Linus, and Bob have their arrows ready for battle.

Yesterday morning it was 58 degrees here which was as cool as I can remember in July. It sprinkled just as we got all the hay in yesterday. Today I expect to attend the cattle-feeders' tour at New Madrid and Portageville and have a dinner at the Delta Research Center in Portageville.

Yesterday we harvested a part load of soybeans even though it sprinkled four times during the day. I voted for four Republicans and thirteen Democrats. The thirteen Democrats had no one running against them here in this area. Only one of the four Republicans was elected.

Linus, I was interested in the list of gas tractors and other equipment at Sycamore, Illinois. There are a number of things on the farm here which could fit such a display, such as a 1934 and 1935 Model A. J. D. [John Deere] tractor and a 22 inch Red River Threshing Machine which I traded in on an automatic J. D. hay baler fifteen or twenty years ago, but the dealer never removed it from the farm.

Tonight I am alone, for Alma and Mrs. Heath are in St. Louis with Ruth or Alice. Mark Edward, Mary's son, is with Jeanie in Poplar Bluff. Mary is on nurse duty with a lady in Lilbourn. We expect Alma, Mrs. Heath, Ruth, and her two sons to be here tomorrow night which should speed things up a lot here.

Last Tuesday night it dropped to 32 degrees here but still didn't kill our tomatoes. We did combine beans five days the past week.

Alice and her husband have purchased a house which they expect to move into in three weeks. It has rained every Monday this month. There is an old saying that when it rains the first Monday in the month, it will rain the next Monday also and that there will be at least fifteen rains during that month; well, we have had at least seven rains now, but I should be used to that after forty years here.

Today Alma will also bring her aunt who lives in Cabool, Missouri, and is now visiting Jeanie in Poplar Bluff. She is the lady who wrote, "Hillbilly Homestead" after she was past seventy years of age. In it, she has a chapter about the Gumbo Swamp and the Blue Flag Ranch as this place was known back more than forty years ago when she had lived here.

I read in the material that David gathered that our grandfather L.G. Hubbard was 5' 6 1/2" when he enlisted in the Civil war; later he was 5' 8' tall and weighed 160 pounds, so I am larger than he. Does anyone know how tall grandfather Koehn (maternal) was? David mentioned also that L.G.H. was told to climb a tree and shoot at Pickett's charge at Gettysburg. I marvel how much of our Hubbard and Koehn history David has learned about in recent years.

March 17, 1971

Last fall Jeanie's white German shepherd had pups, and she gave us one of them. He is now four months old and larger than any dog we have had before. We never let him in the house, but he meets us at the door as we go out, and he is here to greet us as we return from town. If I had half his enthusiasm or energy, I would be all right, for my energy has declined lots in recent years, but my health is good and for five minutes I can do as much as ever. I am happy that he does not bark too much, but only serves as a door bell.

Mary and Mark have fifty-one Austria white pullets which arrived three days ago, and the recent mild weather has been nice to them. Over half the calf crop is here, but I may have fifteen more in the next three months. I have turned most of the cattle out of the lots, because I am running out of hay. Usually I can plan to pasture cattle after the first of April. It has been too wet all winter to pasture winter oats or wheat.

Radishes and lettuce have come up in the garden, and the chickens are bringing up the onion sets too. I planted forty pieces of potatoes between them.

Last Monday I was in Memphis at the annual Farm Equipment Show. I saw the new 7020 John Deere tractor, a four wheel drive tractor that can turn around in seventeen feet.

This morning I have been reading a book by Ralph W. Sockman. I read that scrubwomen seldom, if ever, have nervous breakdowns and that ditch-diggers seldom, if ever, have heart trouble. He says that it is good for us to use our large muscles. I guess that is why I like what I do, for in my daily work most is physical and out-of-doors.

However, yesterday afternoon, I watched Danville and Ohio State lose to an all black Western Kentucky team on T.V. Two days before I saw Ohio State beat Marquette. I am a great T.V. fan and keep up with five or six different sports broadcasts each day when I'm not farming.

Because the Northern blight ruined part of the corn in this country last year, we have no reduced allotments. There is no penalty if we plant the whole farm in cotton, for there are no marketing quotas, but who wants to raise cotton?

July 13, 1971

This seems to be an unusual year in more ways than one. I hardly know what will happen next. Since June 6, Alma has spent more than five weeks in the Sikeston Hospital. First it was a mild heart attack and now a clot in one lung. A week ago today she had surgery near midnight, and as I understand it, they placed a filter in a vein to prevent a clot forming in the other lung. The girls are taking turns staying with Alma. Alice stayed with her Saturday night, and Mary has been with her the last two nights. Jeanie and Ruth take turns staying with her during the daytime. At present Ruth's boys are in Poplar Bluff. Last week they were here.

Tonight the dog has been barking since midnight. I wish he would shut up so I could sleep. I fed him but that only stopped him while he ate.

As usual this is the season when weeds grow even if the beans do not. I pull some, mow some, and cut some down each day but still they come. Foxtail seems to be taking charge in the garden, and swamp grass is six feet tall. I don't know why I should garden any more any how. At least I do have some corn in the garden that is taller than the grass.

To begin with, the garden was planted with a mule planter with yellow sweet corn in one planter box and Truckers Prolific white corn in the other box. We have now eaten most of the sweet corn which was hybrid and only grew four feet tall. The Truckers Prolific open-pollinated white corn is much like the white St. Charles which we grew here forty years ago and is now about fifteen to eighteen feet tall.

Tonight I may see the all star baseball game on T.V., and now I am Mark's night nurse. He coughs quite a bit in the day time, but neither last night nor tonight has he coughed.

October 16, 1971

It hardly seems possible that this month too is half over and that we now have relatives living in many states and even in Canada. Yes, I never heard of Uncle Father-in-law before and to think I have two brothers who are such.

MJS: The son and daughter of brothers Joe and David wed. They were first cousins, hence uncles and fathers-in-law.

Well, in the Bible Joseph's father and mother were first cousins, so it may be only natural for us to think a lot of our relatives. Years ago I gave my best doll to cousin Dorothy Hubbard who died when she was only fifteen.

With dry weather bean harvest in full swing with moisture below 11%, we work early and late. Early wheat is up, but late wheat waits for rain to sprout. Several mornings I could see my breath early, but as yet we have no frost. The ditch in front of the house is bare and dry, and our mosquitoes are no longer around.

October 30, 1971

For two weeks now the Robin has been here and during that time the Illini have won their first football game of the season, and the weather has been fine with enough rain to bring up all winter oats or wheat but not enough to drown out the low spots where I have failed to run proper ditches. We have combined nearly half the beans on sunny days and have had no frosts as yet. Tomatoes are still producing in the garden which I disked in late July, and no volunteer cucumbers are bearing. Corn has tassels and silks on it. There are also volunteer squash and muskmelons with blooms on them and lettuce nearly ready to cut. Wouldn't it be nice if it did not freeze all winter?

Today is the last day before daylight saving time ends. Such time "kids" some people. I always contended that it is for city people and not for me. I suppose I am like a rooster who crows in the night long before daylight and thinks all others should do the same. Is daylight saving time really intended to move people along who need a push?

I seem to be very happy today, but that is not so unusual, for God has certainly blessed us. Alma is almost back like her old self. Thursday she had her club here with ten women. The next day she went to Kennett with Mary and Grandma. Today she intends to go to Poplar Bluff. I expect to stay here and move beans, for it is nice to move them when there is no mud. We waste lots of beans when we see water everywhere when we combine.

February 6, 1972

This morning it rains, and from the sound of it coming down outside, it's not just a drizzle. However, the ditch was at least low yesterday, and even a three-inch rain would only fill it, so I should be happy that it is February and not April.

Our calf crop started January 26, and now I have eleven new ones which is one-third of the total number I expect this year. This should be the smallest calf crop in forty years...

Yesterday we were in Poplar Bluff to consult with our CPA son-in-law about our income tax. Since there was so much flu in the town, we did not go out of the office, for as yet we have been well here and did not want to bring over what they had. Bill said Don is getting to be quite interested in chess and consistently beats his mother now.

January 19, 1974

I have short wheat planted just before thanksgiving, but most of the 200 acres this fall made lots of pasture and was early. It still looks green with all the ice and snow, but last week we had another four inches of rain which ended the pasture... The whole farm was covered with ice last week, and power and phones were out most of one day.

With new daylight saving time, Mark's bus comes in the dark, and he dreads to go. It has been harder to get used to new daylight saving time, and I'm sure you know what I think of this new deal of the Nixon administration. The only bright spot on the national political scene seems to be Henry Kissinger who must be a blessing to the world. Ruth seems happy in her new home and 1974 should be a blessing to all.

February 5, 1975

Most of the sports I see are on T.V. However, Mark and I went with Bill Denton and his family to Gideon where Lilbourn beat Risco in the County Tournament. Risco grabbed the lead and held it most of the first half, and for a time, four of their five players were black although Risco is mostly a white school. They had five players by the name of Johnson on their squad.

A few weeks ago I saw Angela play her oboe on Cape T.V. and last week saw her on T.V. again singing in the junior high chorus.

February 12, 1975

Today is the twin's birthday, and how well I remember their arrival. It was warm and dry then just as it has been in the sixties daytime here with sun the past three days, but then we had snow the last two weeks of February and the usual slush afterwards.

Monday we were in St. Louis to celebrate the February birthdays. The five youngest grandchildren in the family played on Paul's slide in the back yard. They would climb a ten-foot embankment of grass, then slide down on their feet, their pants, or even knees. Once I saw Paul come down end over end, and I thought he might have broken his neck, but he picked himself up and came down again with much laughter.

March 10, 1975

This morning the ground is white outside, for it snowed late yesterday. . . Last week Mark had his hernia fixed in the Sikeston Hospital, and now he walks bent over like an old man, but his spirits are good. A few weeks ago he fell off the monkey bars at school, and he tried to use two canes for crutches for a few days, so he has missed school since daylight saving time came on again, which would have put him on the bus in the dark. However, Alma and Mary teach him at home, and since he is a good student anyhow, I guess he will get along.

Mrs. Heath is now in a nursing home in Poplar Bluff after a trip first to the hospital there. We think she had a stroke.

About a month ago I set out some onion sets in the garden, but the only ones which are up are those the chickens uncovered. Perhaps I will plant some garden on St. Patrick's Day.

(P.S. from Alma, March 1975) *I really have no news here except medical problems of some member of our family, also we look after Nancy & R.T. (he broke his hip, and she is diabetic so must go once a month to a doctor). Most of the time Mary drives, but she was sick with the "flu" for 3 weeks and 3 trips to a doctor, plus Mark had a vomiting virus (it was in the school) and his side was so sore we took him to emergency room at hospital to find if it was appendicitis. Yesterday Mary took Ralph to Sikeston because he drove a post and received an eye full of mud and dirt. The big problem is there are so few rural doctors! We first took my mom to the Sikeston hospital. It was completely filled. We took her to Poplar Bluff hospital, then to a Nursing Home. The stroke was slight, but she can't walk and takes lots of care for there is a bit of confusion. She is improving, but Mary and I can't lift her. She wants to come home and I don't like to leave her, but I have no other choice for a while. Jeanie visits her, and I call her.*

September 20, 1975

Today Illinois plays Missouri, and how happy I will be, for how can I lose this time? I do believe that Illinois' Bob Blackmon showed poor judgement when he didn't take the 55 yard field goal against Iowa when it was a score already made, but anyhow the Illinois team won, so all was fine.

Yesterday we had another two-inch rain. I thought surely by now our two to five inches a week rains would cease, but no dry weather seems in sight. I guess though, I would rather have rain now than in another month when we would like to combine beans... I seem to be having hay fever for the first time in twenty-five years. Even now though it is not as bad as before those six silver clips were placed in my head.

Yesterday was Teresa's tenth birthday, and she is to start playing the flute that Ruth and Alice used. Angela can now play my old clarinet a lot better than I, for I really did not put in the practice she does. When she was here, she blew my old A clarinet, but thought it took more steam to blow than the B-flat one. I admire that Bob both plays his violin and sings at nursing homes and other places.

David writes that he swims a mile a day all summer. I have waded mud in boots nearly a mile three days a week all summer, for it surely has been a wet one. Yesterday after the two-inch rain, the water was again rushing south under the bridge in front of the house like Current River near Arcadia. At least the ditch has run better since the dragline operated in June...

December 26, 1976

As I read the various letters and Christmas greetings, I know that we all have many joys, happiness and perhaps some tribulation, but all are to be expected. At present Alice is here with her three; namely, big John Dan, little Paul Dan and smaller Grant Andrew, who is five months old.

Alice and Dan find the farm entertaining for their sons Paul and Grant.

Yesterday morning Grant was sitting alone in our living room watching Mark and Paul playing with their Christmas presents. He moved along forward more than a yard by merely bouncing himself in a sitting position, waving his arms and showing how happy he was. He cannot crawl yet, but has a stroller which he rides in and thus gets himself up to things he thinks he needs to investigate.

I find at this time of year, I too count it a blessing to be here and enjoy the grandchildren, for my dad never did see any of his. He was born in 1876.

Ruth teaches in Portageville but not in high school, and needs certain qualification if she is to teach in grade school where she prefers to teach.

Joe, you say you cannot read for more than a half hour at

a time, and I am like that too, for I soon tire and get a headache. It's also hard for me to stay in the house more than a few hours at a time without headaches. Yet I can drive a car or tractor outside and not have them.

 Dave, we sure did enjoy your visit here a few days ago. I see you expect two men to take your place when you leave the service. I have recently reread two books from the attic which I found as interesting as when I first read them forty years ago. Have you read, "Heroes in Blue and Gray" by Robert E. Alter?

October 11, 1977
Dear All.

 It was good to hear from you. . . . I am somewhat under par with more hay fever than I have had in 27 years, and although I sound nearly every night like a steam engine, I really run out of steam and have little energy to do those things I might like to do. Perhaps we will have enough dry weather the next sixty days to harvest our crop.

 Today is Paul's birthday, and Alma called Sunday and talked to the little fellow. He talks well over the phone and is very enthusiastic for a five year old. Alice has her hands full with Grant, for there is nothing he overlooks in his investigations.

 Mrs. Heath is now in the Sell's Nursing Home at Matthews which is only twenty-two miles from here. It is a much easier drive from here when Alma sees her and less responsibility for Martha. Since this place is eight miles south of Sikeston, she is not too far from the hospital and doctors. This week she will be eighty-eight and has outlived all three of her sisters.

 Mary is quite involved with a pack of thirty-two Cub Scouts at Lilbourn. This year Mark attends school at Howardville and is in fourth grade. Steve also is in Cub Scouts at Portageville, and Ruth is teaching Home Economics at Gideon, a fourteen mile drive from Portageville. Joe plays little league football.

February 19, 1978

 In a few days now it may be time to make an early planting in the garden, but it snowed another four inches Friday night.

Last summer, Donnie went to Mexico when his band made the trip. Next year he plans to go to school at Cape, for he is eighteen today. Angela will be sixteen in June. Steve has been here this weekend and went with us to Mark's Cub Scout Blue and Gold Banquet. Steve is in Cub Scouts at Portageville, and Mark hopes to go with Steve when they have their banquet.

Mrs. Heath is in a Nursing Home in Matthews and asks Alma to bring her back here. It almost breaks your heart to see. Alma does what she can for her there, for her mind sometimes is very clear, yet she is down to about a hundred twenty pounds.

February 28, 1978
. . .We have recently had warm weather, which had caused enough melting to make the ditch nearly full of water in front of the house.

This winter Ruth has been teaching Home Economics in Gideon. She seems to enjoy it but feels the kids in school there are not wonderful. She said she talked to two boys in the hall enough to stop a fight, which seemed to give her a sense of accomplishment. I think she has some boys in her Home Economics classes too. Last Saturday I talked to a former Lilbourn Agriculture teacher who now has girls in his Vocational Ag classes in Charleston. Some of them do better in FFA than the boys.

November 3, 1978
. . . Yesterday Martha, Mary and Alma were at the nursing home at noon, and for a time they all thought Mrs. Heath was dead. Jeanie and Alma stayed all day, and then came home at 6:00 P.M. and reported she was better. Jeanie then went back to stay with Mrs. Heath last night, but thought she would go home if she improved.

Mark is now in Scouts, although a little too young. He expects to be a tenderfoot by Christmas. . . I have fixed up the (ping-pong) table in the garage where Mark likes to play with me. Mark still likes all games as much as Bob, but at present prefers football. We pass and kick in the yard. I can still kick further than he, but cannot throw over handed any more, so I have to pitch under handed since I am so muscle bound. Steve passes well.

David, It must have been very good to see Ray Dvorak (music teacher, like his father) at the Wisconsin game. Those were good years when he taught at Urbana High.

March 13, 1979

I have just returned from a trip to Urbana where I heard more speeches by more different people at the Agriculture Alumni program on March 10, than I have ever heard in one day. Art Mosher "32" was honored, as was his Dad who was about the first County Agent in both Illinois and Iowa. Since Art's graduation, he has spent most of his time in Indiana teaching them to produce food. It was also thrilling to see what's new in electricity. Brother David and I looked in on E.E. [Electrical Engineering] open house where I had not been in the past fifty years. Mark would really have enjoyed it if he could have been with us there.

We had seen lots of water near Cairo and for ten miles north when we crossed the new bridge near Charleston on I 57. As we returned it looked very high yet, but they are not using or expecting to use the Cairo-New Madrid Spillway as they did in 1937.

On Sunday P.M., March 4, we saw Don Stewart sing in a beautiful presentation of "H.M.S. Pinafore" at Cape Girardeau. We also were in Memphis at Cook Convention Hall at the Farm Show there on March 3.

January 13, 1980

Another year has rolled around, and now I've reached the ripe old age of seventy. I would like to live another seven or eight years, but realize I can't expect to live on forever, even though I have been most fortunate to still navigate with health as good as many others.

With the end of basketball season I feel that at last spring has sprung. I don't think I have ever seen as many good college teams play on T.V. before in my life. I was surprised to *notice that the Illini has as many black players as Lilbourn and Charleston who won the state title.*

March 29, 1980

Yesterday Jeanie brought Teresa here to stay over night. She is now as tall as Angela and plays the flute very well. I was surprised how well she could transpose and play with Mark. She too is a freshman... Mark plays the tenor B-flat saxophone. Donnie Stewart has gone to Denver during this Easter break (to ski). It may have snowed so much, he may find it hard to return.

We had a big rain here, and twice yesterday I had to poke drifts which had collected against the north side of the bridge where Heaths used to live, so that it would go on south to Little River and of course on to within three miles of Gideon and Kennett. The lowest point in Missouri, of course, is about twenty-five miles south of Kennett. All our water goes 200 miles south before it ever enters the Mississippi.

We watched Mark play basketball Saturday morning with a bunch of other sixth graders, and he managed to make a basket.

March 31, 1980

Tomorrow is April 1 and election day. Since most schools in the county now are consolidated, we vote here for building a new grade school in Portageville. Joe and Steve were transported from Portageville to Conran for fourth and fifth grades.

While Dan's father was here, he mentioned that while the family grew up in New Hampshire, his brother visited in Greece, and since he was born there, was put in the Greek Army for two years while they were fighting Mussolini in Albania. Since that time, he lives in Australia and was here when Dan and Alice were married.

I have set out twenty-five "Sure Crop" strawberry plants which I ordered from Iowa. They were pretty well dried up when they arrived. I set out five pounds of onion sets on Washington's birthday and also have some lettuce and radishes up. The magnolia is about to bloom in the garden, as are the early flowers around the house. It has been too wet to do any disking since last fall.

When I plant, I spade up an area near enough to the pump house; then I break up the clods and plant the things I think should be in the ground early for best results. For a while I had an urge to try some dewberries, since people pick them wild on the railroad tracks near the washout north of Lilbourn. They are bigger and better than blackberries.

July 27, 1980

Tonight it started raining at 12:25 P.M., so I hope the drought has broken. It has been the driest year here since 1936. Of course, we have had other years when two-inch cracks appeared between all rows of the crops that were not irrigated.

The big elm south of the house died as a result of the dry weather. We cut it and pulled it south with chains so that it would be sure to fall away from the house. Dan has been sawing it up to use in his fireplace this winter. Yesterday afternoon Ruth brought her boys here from Portageville, so five grandsons watched the operation. Alice's boys have been here most of the past six weeks.

A few weeks ago Mark went to Scout Camp a hundred miles northwest of here. He also attended the Methodist youth camp at Arcadia in early June with David and others from Portageville.

Dan and I took in a part of the Governor's Spotlight Tour for Southeast Missouri. I ate with about 100 other farmers at New Madrid last Tuesday but did not attend the fish fry at Sikeston that night when 800 people and the Governor of Missouri was present.

The peaches at the big packing plant west of Malden are no larger than walnuts this year.

Most of this county now will be going to high school east of the Indian Mound between Lilbourn and New Madrid. They expect to have football there this fall. Portageville will still have their high school, and Joe hopes to be a tight end this year on the football team. Mark will go to junior high at Lilbourn instead of Howardville.

As we get older, I get slower and have a lot less desire to do anything in a hurry.

July 30, 1980

We did have a one-inch rain three days ago, but it is hot and dry again. I find the pictures in this robin quite interesting but cannot imagine being served a pig on a platter with its head on and an apple in its mouth. . . .

Our dog, mostly white with black spots, is very happy when people drive up. She gets clear up on the top of cars and lies down and, of course, smears the windshield and trunk in the process. When I go out to weed beans, she goes out too and picks up weeds which I pull up. This has been a good year for button weeds to appear in the beans, but when I pull them, I usually pull ragweeds or cockleburs too. Although there are no mature cockleburs, the dog leaves them alone.

It seems the Olympic boycott has given the Russians and East Germans a chance to have all the gold and silver medals. I suppose the Americans might have made the track and field championship more competitive. However, what would have happened if the Russians had decided to hold them hostage after they got there?

Now Mary and Mark have returned to St. Louis with Dan and his boys, so for the first time this summer, Alma and I were alone when Jeanie drove over yesterday noon. Today she (Jeanie) planned to go to Cape Girardeau where Angela is expected to compete in a swimming meet. Angel has been a lifeguard at the pool the past two summers.

October 14, 1980

Harvest is progressing fast now and moisture down to 11.5% on some of the beans. The beans are only about one-fourth the size they usually are. Much of the land is going into wheat around here to get away from a possible drought next year.

I have been enjoying the S.S. [Sunday School] lessons this quarter for they have been favorite passages. I have always liked Micah 6:5 and John 3:16, but know many more that are even better. I marvel how many Dad [Fred C. Hubbard] knew and remembered the time our pastor asked him to quote Psalm 1 which I later memorized too.

December 27, 1980

Well tomorrow I will be seventy-one, and I, like the hostages, am still here. Dan, Alice and their two boys came by here Christmas Eve in a big rented van, and now Alma, Mary, and Mark are on their way to Florida with them. I was invited to go too, but after the trip I made to Urbana in Jeanie's van, I felt I should stay here.

Linus, I have a radio that is stuck on KWOC Poplar Bluff. Just now someone was singing a song which said, "You don't see many mules these days, you don't see many mules and Georgia Mules and country boys are fading fast away." I enjoy radios more than T.V. for radios do not give me a headache. If I watch T.V. very long I use colored glasses to prevent headaches.

Alice and Dan left their small dog here. He seems to like the farm and our dog, so now I feed two dogs and three cats.

Someone gave me one pint of Pure Maple Syrup from Newport, Vermont, in a tin container with pictures of people gathering sap from trees. I suppose some Hubbards around Wilmington still do this.

I enjoyed the phone calls I had on my birthday. Ruth had a birthday cake at their house Sunday with 71 candles on it. Steve helped me blow them out, for I was about to blow out my teeth when I tried.

For years I have lived one year at a time. As my activities slow more, I'm glad God has been so good as to let me live this long, and while I no longer run, and get stiff easily if I sit or lie down, at least I don't limp when I walk. I digest stuff much better when I eat than I did when I was in college.

March 6, 1981

It amazes me how time has brought us brothers together. My two little brothers seem now about my age and perhaps are really far more experienced than I, for their problems have not been just like mine.

I was surprised to learn that Joe now goes to a Baptist church not much larger than ours here. When I first came here, *we went to a Baptist revival at the school house at Woodrow where Alma went to school during her early years here. She must have done some walking to dodge the cattle and hogs on the farm to get there. They had a brush arbor revival there in 1932, held by two young men from Risco, and there were 45 people baptized as a result of it. When I first came here, there was a General Baptist Church about five miles northwest of Lilbourn which had their foot washings from time to time. It was there that I got acquainted with shaped notes and Stamps Baxter tunes which I enjoyed very much.*

As I read your very inspiring letters, I realize that we may be starting a very important time, namely, the last days that we heard so much about when we were children. I heard over my radio that another attempt was made to hijack a plane in California, but this time FBI men were in time and made the right moves.

Alma expects to move on to Columbia to see Angela honored at a tea today. Angela's advisors now have her as a part of the Agriculture School, but she still expects to come out a dentist. I do not understand why Agriculture Economics and Animal Husbandry should be part of the training for a dentist, but I was pleased that her advisor was from the University of Illinois and recommended it to her.

MJS: By following the agricultural track in chemistry for vet school, Angela would be prepared to enter med school and pick up scholarships that were not available in Arts and Science.

Linus, I do not know a thing about gout, for no one has accused me of having that. It must be so old, it's new. I am glad to learn there are so many churches in Tifton. There have been all kinds of churches here too. One time I attended two baptizings in Little River within an hour of each other. The Missionary Baptists who followed the General Baptists said their's was the true [baptism] and covered one, for their baptism came from

John the Baptist and was the only true one.

July 1, 1981
. . .Tomorrow Alice's family is to arrive here soon to celebrate Grant's fifth birthday. Ruth and her boys were here yesterday. Steve mowed the lawn, and Joe helped me operate on a tractor and remove a branch which had been blown down off the oak tree near the chicken house.

We had more rains since we came home. It took more disking this year to disk in the wheat straw, for much of the wheat made close to sixty bushels but brought $3 per bushel which was a dollar less than for several years.

Ruth and David's boys find working on the farm a real adventure.

July 12, 1981
Last night about 7:00 P.M., Don Stewart arrived here after spending a month in south Mississippi near the ocean. He showed us all kinds of sea life he had collected while he was there. Many of them were edible, but didn't look so to me. After a brief stay in Poplar Bluff, he plans to take a train to Mexico where he plans to spend the next month, after which he plans to be back in Cape Girardeau for more studies. Angela is going to Three Rivers College this summer and making credits to save taking them in Columbia (at Missouri University).

Yesterday we were in Dexter to see Joe, Teresa and Mark march in a big band contest just as we used to participate in band contests when I was in the U.H.S. There were bands from as far north as Farmington and Potosi and as far east as East Prairie and Caruthersville. The hills in the west part of Dexter were just covered up with bands with 1,000 or more kids. There were bands which made you think of the American Revolution and one band from Richland that had uniforms like the Southern Confederates. Teresa wore a maroon uniform with a white cowboy hat and white gloves; Mark wore orange and black while Joe wore mostly bright blue. Mark keeps step nicely in his fifty-four piece band.

November 2, 1981

Mary and I returned yesterday from a trip to Urbana to watch the Illini beat Iowa where I met many friends of the class of '31 at the fiftieth reunion. We ate dinner in the Champaign Country Club just before the football game. Lee Sentman of Decatur joined us. I remember when he was a one-man track team in the hurdle events and the broad jump both when he was in high school and in college. David and I played chess each morning before sunrise, but he won most of them.

Grant rode with me as I combined beans yesterday. A five-year old has a way of getting things done. Paul and Grant played in the truck load of beans, and Mark and Steve did the same thing today. Paul and Grant rode lawn mowers, so most of the leaves are crushed into small bits so the wind can blow them away... Teresa, Grant and Paul attempted to ride their bikes, but the road was so muddy the wheels filled with mud.

Riding a lawn mower on a flat farm makes Paul feel safe.

June 27, 1982

This morning as I write it sounds like a mosquito is singing in my ears. We have not really had many of them, and it has rained very little the last month although it showered some last night.

Linus, you report a tough winter on your foot. My feet don't stand a whole lot of walking without hurting afterward.

Mary, Mark, and Steve (Ruth's boy) returned after Mark had spent a week at Scout Camp, and Mary and Steve had fished with Bill near Poplar Bluff.

Last week Alice, Dan, and their three sons were here two nights for a birthday party... The new boy, Matthew Kent, is a really good baby and now is fair and bald headed with brown hair near his ears.

It is quite different now than when every one was reporting snow when I last wrote. Now people are selling fire works in a stand between towns to celebrate the Fourth. Fifty-one years ago they sold the fire works mostly at Christmas time. The robins that usually come through here two months ago as they go north are still here, but our bird sings less this spring. Perhaps that might be because the tall sycamore tree that stood on the ditch bank was cut last fall. It is very hard to use either sycamore wood or elm wood for burning since no one can split the wood after it is cut.

Yesterday we saw a pheasant in the garden. This is the first time I have seen any of them this close to the house since I raised twenty-five of them about twenty years ago. I got the chicks from Tennessee. It was in June, and I raised them along with other chicks. When they were about six weeks old, I turned them loose and most of them went a quarter mile west of here to the fence row between here and Dentons.

Now most of the wheat has been harvested and planted back in beans.

Joe Richardson has been working on the house, painting the windows and trim. Teresa Stewart works some at a Bonanza Steak House in Poplar Bluff. Angel Stewart works at a hospital and is assistant coach for the Poplar Bluff Swim Team. Don Stewart is attending Washington University in St. Louis. Angel is now a member of Alpha Zeta. I didn't get in A. Z. till I was a junior. In 1930 I can remember no girls in A. Z., also no Blacks.

Our church was having a picnic in the park in the pavilion when it started to rain. There certainly was a lot of thunder at 12:30 p.m. as the rain began. The children used the slides in the park after the rain, and a three-year old landed in the pool at the end of the slide when her dad failed to catch her at the bottom of the slide. The funny thing about it was her disappointment that she could not continue her sliding even though she did get really dirty when she hit the end.

Dave, for the first time in my life, I seem to no longer desire to rise early or do much. While I've raised a garden, I am happy to let others harvest it.

November 24, 1982   Thanksgiving
Dear Relatives,

As you see, today is turkey day and Alma is expecting eighteen here by noon. It is cold outside, perhaps 20 degrees but no snow or rain. We did have several drizzly days during the past two weeks, and now there is two feet of water in the ditch for the first time in a long while.

I see the Big Ten will be represented in five bowl games this year. I suppose that's the first time that has ever happened.

We do have a lot to be thankful for. We do all walk around on two feet, and while we may moan and groan, we still get around.
(Sunday, 11-28-82)

Now we really celebrated Thanksgiving with a big feed here and another the next day in Poplar Bluff. David Richardson christened Kent Matthew in Portageville, and Bill Stewart took a picture of all of us in front of the church. While all of this was happening, it has rained the past three days. Paul and Grant are surprised how much water there now is in the ditch, but after two weeks of rain what should one expect? Lots of times it rains for two weeks and then is dry for about that long.

I have seen so much of T.V. football and basketball the past three days that I am sure that I can wait to see the Illini collide with Alabama on Dec. 29.

Alice's baby Kent is the most interesting creation. He smiles a lot and now has a walker which he propels about the room and up to things he wishes to investigate; he rolls up on all fours and looks as if he is going to crawl but does not crawl yet. Sometimes he moves by rolling when he is not in the

We celebrate our last Thanksgiving.

walker. *The other day he rolled about ten feet to where I sat and then reached up for attention.*

Joe, Mark and Don are all near six feet tall. Mark now wrestles after school and had a meet which they won at Charleston. Next week they may go to Poplar Bluff. Joe's football season is over, and he plays basketball. Even Paul and Grant play some basketball at school, although Grant is only six.

I have enjoyed all of your letters, and while we are not as we once were, we still can look to the future as well as the past. This little one which Alice has brought here makes me think of things she did when she was his size, for she too was quite a do-er.

Joe, I realize that winter means you stay home as I do. I seem to care less and less whether I go or not. I was glad I attended the 55th reunion of our U.H.S. class of 27. I always see some one whom I could not remember. This time I saw Wellington Towner and Bob Mason. On our wedding trip nearly

50 years ago, we saw Margueret Towner and Harmon Roberts in Indiana at "The Shades" as we were on our way to Urbana to Bob and Faye's wedding. We all went swimming there in some rented swimming suits, and they took our pictures.

Bob, we all pray for you and appreciate your life as an older brother who sure took care of us younger ones. We have been so blessed in growing up together in a very happy home.

With love, Ralph
Written by Ralph in November 1982

Ralph and Alma celebrate their Golden Wedding Anniversary on June 4, 1983 at Kentucky Lake.

Flowers celebrate a special anniversary.

The celebration of Ralph and Alma's 50th Wedding Anniversary was a family affair because Ralph's stamina was waning. Never did he complain, but no longer did he have the energy level to enjoy festivities. The family spent a week-end in a lodge at Kentucky Lake where children could play and adults could rest as needed. Flowers and gifts were presented. Shortly after returning home, Ralph entered the Sikeston Hospital with a bladder infection, developed pneumonia, and died July 6, 1983.

A Goebal figurine of Mary and Joseph was presented to Ralph and Alma along with a note of appreciation signed by all the children including the thumb print of baby Kent.

A Golden Anniversary
June 4, 1983

We want you to have something to remind you of this special day. This gift speaks of love--God's love for us and our love for God; man's love for a child and his betrothed; and a mother's love for a child and husband. It is the same love you have given to each other and to us--steadfast, true, human, and divine. No gift can be greather than the gift you have given us of your love that has lasted fifty years. It is the same love we see in Joseph who faithfully forges ahead with a lantern, walking according to God's leading, and walking so that Mary can ride. It is the same love we see in Mary as she follows Joseph and seeks God's will in all things. In both figures there are lines of hardships and uncertainties, but what really comes through is an abiding love. Because you have loved, you have taught us how to love through all adversity, and that has made all the difference.

Your Kids,

Chapter 7.
Wrapped Up in Farm Life

Living on a farm in the years before electricity, telephones, or television meant that life took place in a closed community. Tasks were handled with simplicity, dependent on the brawn of men and animals to meet farming tasks. As technology crept into the farming picture, farmers struggled with new resources and changed their view of possibilities.

In the letters of Alma and Ralph Hubbard, we walked with them as they comment on every day, mundane happenings within their family. We listened as they attempt to put into perspective world events, to encounter the trauma of World War II, and to seek to raise the standard of living for the poorest of people. We felt their pain at encountering the injustices of segregation. We saw them wrestle with establishing a public school system and found in their religious faith a force that sustained their lives.

Our farm was a bustling village of eleven tenant houses in addition to our home and the home of my grandparents, Joe and Mattie Heath. We were a self-contained community, interacting with one another through work and play, supporting one another emotionally and monetarily, sharing pleasures and troubles, and always in the business of educating one another with "what works best." As a twin I came into the world with a ready-made playmate. Mary is the one who walked with me through all of life's encounters good and bad, who listened to my suggestions, and waited for me to solve our problems. I felt responsible for her all my life.

Life and birth are facts of life that are easily assimilated on a farm. Chickens set on eggs. Eggs hatch. The feeble, wet chicks

Ralph brought kittens from the barns to Alma for cat care.

emerge and struggle to stand alone. Soon they are scampering about and acting like chickens. Cats have kittens in the barns, and Dad brings them to the house. The kittens nurse, kneading their mother with their tiny paws. We stroke their full tummies and are ecstatic when their eyes open and blink at their first vision of a new world. We, too, blink at the world of possibilities.

At four years of age, I experienced my first birth. It was not something that Dad and Mother set up as a learning experience. My sister and I were exploring our spacious fenced-in yard, when an interesting event captured our attention. A sow and her baby pigs came across the adjoining pasture to our yard's fence. She plopped down to let her pigs nurse. What entertainment! Mary and I plastered our bodies against the wire fence and watched the piglets fight for dinner as the sow snored and grunted contentedly. Then it happened. First some bloody fluid oozed from the sow's hind quarters. Then, with a heave, out came a gush of fluid and a translucent pig. Immediately the squirming baby pigs shifted positions and trampled it into the mud. We were puzzled about the thing that looked like a pig without any hair, full of bloody veins. When Dad came later, he remarked that a dead pig had been trampled and wondered if we had seen it born. Indeed, we had!

We were rather self-sufficient at producing what we needed. A large garden kept us in fresh vegetables and provided extra for canning. Asparagus plants brought from the University of Illinois thrived in our rich soil and furnished asparagus each year until the birds seeded it in all the surrounding fence rows, and then we cut asparagus in the wild. We had our own orchard with apples, pears, plums, peaches and cherries. A grape arbor boasted different varieties of grapes. Gathering eggs and feeding the chickens was a chore delegated to the children. In the chicken yard, the meat walked around until we were ready to eat it – nature's alternative to refrigeration.

Every morning and every evening, Dad would milk the cow in the pasture behind the house, no matter how tired he was. He offered a cow to his workmen, but usually they did not want to be bothered with the tediousness of milking morning and night without taking a day off. We always had an abundance of milk products, but we always had milk buckets and crocks to wash. They were wearisome to clean and sterilize. I learned about sterilization at an early age. The mysterious process of pouring a teakettle of boiling water over a milk bucket and crocks kept bacteria from forming, a perplexing evil. Our folks had a gasoline refrigerator on the back porch to keep milk cool, a real luxury. My grandparents had a pump box on their back porch that they used to cool their crocks of milk. It was a home-made creek that took place of the one they left in Wayne County. I remember the praise that was lavished on me when I pumped the wooden, five-foot box full of cool water. The cream on top of the cooled milk was skimmed off and collected to make butter.

Churning was one of the first jobs a child was asked to do. Glass churns with wooden paddles sloshed the cream as we turned the handle that was attached to a set of gears. It was like turning the handle of a toy. Mary and I would take turns. We could see the butter when it started gathering in the cream. We had accomplished something important. Then we watched Mother put the pats of butter into a wooden butter mold and turned them out with a floral design on top. We looked forward to the day when we could make pats of butter with flowers etched on them and sell them at the grocery store. The leftover clotted or clabbered milk was used for cottage cheese. Any leftover milk, apple peelings, or garbage was devoured by the chickens or hogs as a delicacy. It was great fun to watch the recycling that went on between greedy animals. Oh how they would press against the fence in eager anticipation of the tidbits we threw over the fence.

Catron, the village three miles north of us, was an important extension of our environment. I thought it was an exciting metropolis, but I realize it had two stores, a blacksmith's shop, a gas station, a school, a church and four cotton gins. At Catron we got those things we could not raise, like cans of salmon, salt, flour, and sugar. At times we would trade eggs or butter for groceries. I remember watching the clerk in Kell's Grocery getting the things we needed, writing out what we had purchased on a pad of paper, and adding the figures. Adding was a mysterious skill I wanted to learn. No calculators in those days! We carried out what we had purchased in a paper sack. What treasures!

Catron supported the farming community with its cotton gins during the cotton season. The gins really hummed and ran full blast night and day. The cotton wagons would line up so the cotton could be sucked up into the gins. A pipe, much like a vacuum cleaner, would be used to empty the wagons. The cotton rolled through the mechanical arms which extracted the seeds and debris. It was then bundled into bales. Farmers were eager to see how many bales an acre their cotton made. Seeds were used to make livestock food or cotton seed oil. Cotton gins were dangerous contraptions, and each year someone got caught in one and was hurt, even killed. We listened to adults discuss the mishap and felt troubled. The sides of the road would be white with loose cotton that fell from the wagons on their way to the gin, making the sides of the road look as if snow had fallen in summer.

Two days a week a grocery truck from Catron made the rounds of the country roads. Shelves were improvised in the back of an old paneled truck and stocked with all kinds of goodies, or so it seemed to me. Especially did I like the candy they had; occasionally Mary and I got a piece, but most of the time it was said to "rot our teeth." The ice truck was another fixture of the age. The ice truck loaded with blocks of ice would drive up and down the road looking for customers, people who would purchase a block of ice. It would be loaded into a container with big ice tongs and would be placed in ice-boxes to keep the butter and milk cold. An ice-box with wooden shelves was lined with tin and had a drip drawer in the bottom to catch the melting water from the ice.

My parents spoke in hushed voices about the "other" store in Catron—the Delta Realty Company Store. It was a general store offering dry goods and hardware, as well as groceries. My folks preferred not to shop in the "Company Store." The plantation system was a part of the economy of Catron after most of the arable land was placed in cotton. People bought the things they needed including groceries with scrip. That was the way the laborers were paid, a system of bondage. The African American laborers seldom saw money, only "script." They were never able to pick enough cotton to satisfy their debts at the "Company

The mules did the pulling, but the workman held the plow in the hard gumbo.

Store." The system called for laborers to sharecrop. That meant they provided the labor while the owner provided the land, tools, and seed. At harvest time the laborers collected a percentage of the crop. If they managed well, they could get so they would rent the land. Then they bought their tools and furnished the labor while the owner provided the land. From being a good renter, they moved on to being a landowner.

My father preferred to farm with day labor. He did try some sharecroppers, but learned that the people were often not able to plan ahead, and they could not afford to take on the risk of a poor harvest. Hands did odd jobs in the winter, for always there were winter up-keep chores to do on a farm. Each week, hands would get a check or an advance. Sometimes a hand would stay during the winter and leave for another farm at planting time. That was a risk Ralph preferred to take. The pastures were always greener on other farms. Before the days of Social Security, Dad took care of the people who worked for him. He loaned them money and took them to the doctor. He used white workers, for he had no experience with the African American workers that came up from Mississippi and Arkansas. Many times I have seen my mother furnish eggs to field laborers, for we always had more than we needed. To share with the needy was a lesson I learned early before days of Social Security. When the Social Security check became a part of our government system, I remember one man refusing to take the check because he "didn't need no welfare."

Everywhere we turned on the farm, we found exciting ways to unravel life. I remember my first anatomy lesson. Mother lectured as she dressed a chicken. So fascinating was it that my sister Mary and I were ready on either side of Mother at the kitchen sink whenever she prepared a chicken for dinner.

We waited as someone got a bent wire, hooked the leg of a chicken, seized its body, and made ready to pull off its head. My mother never could get up the courage to pull off a chicken's head. That was Dad's job. The chicken's head was placed under the handle of a hoe. Dad stepped with both feet on the hoe's handle and yanked the head off with a quick pull on the chicken's legs. The chicken flopped convulsively on the ground. When the dead bird was still, it was placed in a bucket, and a teakettle of scalding water was poured over it. When the feathers of the chicken were drenched thoroughly, the bird was cooled so that the feathers could be plucked. The hot water loosened the feathers. After the wet feathers came off, the hair on the bird was singed. Singeing the chicken involved holding it over the heat of a burner on the stove until the hairs curled up and disappeared.

Then Mary and I were ready for the anatomy lesson to begin. First Mother pulled out the crop of the chicken, and we examined the contents to see what the chicken had eaten. The crop is the bag under the beak that holds the food before it goes to the gizzard to be ground up, for a chicken has no teeth. Next we watched her disjoint the legs and wings with a knife and cut them into serving pieces. We checked the gristle or cartilage covering the ends of the bones so that they could bend. We saw where the wing feathers sprouted and looked for the oil sack on the tail where the feathers could be lubricated with the bird's beak, dispelling rain.

Now we were ready for the fun part, an examination of the innards. An incision on the lower part of the abdomen allowed Mother to pull out all that was inside. Out came the lumps of yellow fat, the coils of intestines, the heart, the gizzard, and the egg sack with the mass of developing eggs. We watched as she opened the gizzard and examined the contents. Once we found a dime that the chicken had eaten. The figurehead on it was polished to obliteration. A supply of pebbles in the gizzard served as abrasive material to grind up the food particles. The lining of the gizzard was pealed off to leave a clean piece of meat. The egg sack always fascinated us. Sometimes a fully developed egg would be in the sack ready to be laid, along with an aggregate of yolks of various sizes for further development. She cut the heart open and lectured on its chambers. When she opened the chicken's body, we located the kidneys in the back and removed them along with the spongy lungs. When the body of the chicken was cut into serving pieces, our lesson for the day was complete. Later I, too, learned to dress a chicken. Why they called it dressing a chicken, I'll never know!

From the idiosyncrasies of the various farmhands and their families we learned about life as we listened to my mother and father discuss their problems. Two deaf mutes who lived on the farm, Ermin and Bobby, ages nine and six, enlivened our lives with their uniqueness and adaptability. As they grew, so did we. Our house was the first stop on the circuit that the two little guys made in their daily exploration of the farm. Dad agonized about their safety around the farm implements and the busy work of the farm hands. Their mother did not seem to worry about their excursions. They were ready to go to the fields in the wagons, climb into barns, explore mud holes, and help with any interesting activity. If we were busy, we would send them home, but at other times we would play with them. They had learned to "Ya-ya-ya" to get our attention, and then show us what they wished. We learned that they watched us closely, so we could tell them through motions what we wanted. To identify people, they picked a gesture or a trait and had most of the people on the farm marked. My father's identity was a sidewise smile. For an aunt who was large, they showed her big stomach. An uncle was cross-eyed, and so went the identification.

One day we turned up with itchy bumps between our fingers. Mother was puzzled until she realized that Ermin and Bobby had been using our swings. It was the itch. They were none too clean, and the ropes were a good place to harbor the mites that were the culprit. "The itch" was news to me. According to the dictionary, it is a "contagious skin disease causing small, watery pustules in the epidermis." We got rid of the mites, and our bumps went away – after a lesson on hand washing.

Jerry, our pet Shetland pony's bridle and saddle were stowed away in our garage. It wasn't long until Ermin and Bobby slipped into the garage, got the bridle and headed off to the pasture after Jerry. They were fearless.

My mother and father worried about Ermin and Bobby's lack of schooling. There was a school for the deaf in Fulton, Missouri, about three hundred miles from our home. Ben, their father, wasn't concerned, for he got along well without an education. He was an outdoor person and spent his extra time hunting and trapping. At first Ben would not listen to my father's pleas that Ermin and Bobby be allowed to go to a boarding school. Finally Dad arranged to take Ben to Fulton to see the school. Ben was impressed by what he saw. I am sure the sales pitch that Dad delivered on the way prepared Ben to like the school. Plans were made for Ermin and Bobby to go to school. I remember how excited we were when they came home for Thanksgiving. They sat on our porch and performed for us. Ermin counted. He arched his tongue and rolled it against his teeth for "one." For "two" he puckered his lips and came out with a sound like "oo." We were thrilled with the progress he had made, and we had a look at what can be accomplished with special education. At times I remember my parents agonizing about the children's deafness and their runny ears. They were not born deaf, but became deaf when ear infections were left untreated.

Medical help was limited. People relied on home remedies and helped one another. The doctor made house calls. Mary, Ruth and I were all home deliveries. When twins were born on the adjoining farm, I remember mother worrying about them, and Dad made a trip to check on them. He found the grandparents feeding the babies 'tater soup and bean soup with a spoon. "We just watered down the beans and made some soup, and watered down potatoes and made more soup." They had a bottle of orange "sodie" that they were also offering the babies. Fortunately, the babies thrived. The invincible nature of humans lies in our genes.

Many families came and went on the farm. Nancy and R.T. came and stayed. Nancy grew up in the backwoods of the sawmill in a tent without a floor. She had no education and signed her name with an "X", but was willing to work. She married R.T. who was like Silas in "The Death of the Hired Man"
(Robert Frost: "Silas is what he is – we wouldn't mind him – but just the kind that kinsfolk can't abide.")

Nancy could chop an acre of cotton a day. She helped mother on wash day and with ironing. It was the day before drip-dry fabrics. Everything was hung on a line outdoors to dry, brought in, sprinkled with warm water, folded tightly, and ironed the following day. Nancy did not do a very good job of ironing but was fast. Mother would redo the better clothes. Her body odor was pungent, for this was in the days before running water and bathrooms. All tenant houses had outhouses. Nancy ironed in the back bedroom, and we aired it out after she left. I learned to iron by tackling the bundle of handkerchiefs that was left in the bottom of the ironing basket, for it was a day before Kleenex.

R.T. could not speak plainly and was somewhat handicapped physically, but was always pleasant. He loved flowers. He potted flowers in tin cans, battered buckets, and broken crocks. He planted flowers around his house, in the yard, and in the garden. His house stood by the feeding barn where he and Nancy could raise a prize garden and keep an eye on the livestock. R. T. would help Mother with yard work, moving plants, filling flower beds or setting out bulbs. On Sunday mornings he put on a white shirt, took his Bible and started walking to Catron to church hoping someone would give him a ride. Dad watched out for him because R.T. would try to work too hard and get too hot. He was not strong enough to drive mules, nor mechanical enough to drive a tractor. He could pick cotton or help shuck corn. From people like R.T. and Nancy, I learned to look for their contributions to a simple life and see the compassion that my parents practiced in dealing with others' inadequacies. Never did we laugh at someone's ineptness. Nancy swept the dirt into the corners of her rooms. Her pet chicken lived in the house with her. After all, it was cleaner than a tent without a floor. Their daughter, Florence, was our playmate and a high school graduate. She and we learned the value of an education.

An essential part of a farm education comes from the interaction of humans and animals. In the early days, the mule barn ruled the farm. The day's work started when the harnesses were put on the animals and ended when they were taken off. Workmen learned to judge the health of the animals as well as their personalities. Mules all look alike to me, but not to the farm hands who work them daily. In 1933, twenty mules were housed in the mule barn on our farm. Unlike tractors, they had character. They were paired for compatibility – Blue and Maud; George and Nig; Fred and Mike; Big Blue and Kate; May and Slim; Pete and Sally. Some were cantankerous, some docile, but most were stubborn. So it became a game to outwit them. The mules would tower over me and intimidate me, pulling their ears down in what looked like a frown. Dad said, "If you want to see what the world looks like, crawl up on the back of a mule. You can see for miles."

The men who shucked the corn used mules to pull the wagons through the fields. The mules wore blinders on their bridles so they would keep their attention on the task at hand. A tall sideboard was added across one side of the wagon. Men took the husks off the corn and slung the ears into the wagon, using the tall backboard as a basketball player hits the backboard in basketball. They gave their teams oral commands and walked through the cornfields husking corn. Mules soon learned the commands to "gee" (turn right), "haw" (turn left), "whoa" (stop), and "get-up" (go). Verbal commands were a part of the game of life that even the simplest of animals learned.

I loved to watch the men unload the wagons when they came in at noon or in the evening. I would call, "The wagons are coming. The wagons are coming," and Mary and I would race to the scales at the big crib near our home. The wagons would pull onto the scales to be weighed with a full load of corn.

Mules ran an escalator to dump the corn into the crib.

After weighing, a pair of mules was hitched to a center pivot and marched around in a circle while a series of gears propelled the corn up an escalator to the top of the crib and dumped it in a vacant bin. The empty wagon with its pounds of gumbo was weighed back through so the men could be paid according to the amount of corn they picked.

Perhaps the most interesting mule on the farm was Roamy, who could scale any gate or fence and roam at his pleasure. Roamy discovered Jerry, the little Shetland pony who lived in a shed in the pasture behind our house. That was the end of any pretence of controlling Roamy. Jerry loved the companionship of the big mule, at least three times as tall as he. Together they moseyed around the pasture – big Roamy and little Jerry – munching grass and socializing. Jerry was a purchase that Dad made at the sale-barn, a worn out animal from carnival life. He was quite docile, but quickly learned that farm life was for him. Plenty of food turned him into a frisky contender for the good life, and after a month or so of delectable pasture, Jerry would not accept a saddle without protest. Mary and I were afraid of him and would ride only as long as Dad would lead him. A bicycle was much more predictable and did not kick!

In 1939, Molly Brown joined the coterie of animals. She was a high-class, gaited horse with a white star on her forehead. The luxury of having her own stall set her apart from the mules. She served as Dad's Lincoln Continental for travel to the center of the farm traversing all the mud puddles in grand style. Her slender hoofs could cut through the gumbo efficiently. No longer did Dad have to walk to the center of the farm in rainy weather to prime the pumps and start the gas motors that ran the pumps at the big stock tanks. At some point Dad put fish in the tanks and entertained the grandchildren in Simple-Simon fashion with fishing for the goldfish that multiplied. There were two concrete tanks at the livestock barns in the center of the farm, one at the mule barn, and one at the feeding lot. The grandchildren fished in all of them.

In many homes pets are an integral part of the family. Pets sit by the table during family meals, go for rides in the trucks, sleep under the bed at night, and today are even housed in motels on family vacations. Not so, on a farm! Animals surrounded us—in the barns, in the pasture, in the wild oats, on the ditch dumps, and in the fence rows. Baby chicks arrived every spring. A venturesome hen who had hidden away somewhere in the barn would prance forth proudly with a string of chicks in tow, clucking musically as she scratched in the dirt to find hidden treasures worth eating. We would sprinkle bread crumbs for the chicks and found their scurrying around our bare feet delightful.

Baby pigs appeared behind our wood stove in a beat-up tub. The warmth of the stove and a dirty gunny sack comforted them, for their mother had died giving birth. We delighted in helping Dad feed them with a bottle. A mother cow and her calf, which needed special nursing care, appeared in the pasture behind our house. We watched as Dad gave the mother and her calf first aid. Always there was a litter of kittens that appeared magically in our garage. The barn cat needed help with her kittens. We held the kittens, played with them, and fed them while their mother purred contentedly. A mother skunk and her three babies entertained us by coming in our yard each night and eating clover in the moonlight. You can be sure there was no hunting for lightning bugs when they came around.

Serendipity played a part in our animal game. One day Mary and I found a six-foot wire pen in our front yard, and in the pen were two black and white bunny rabbits. We learned that by lifting the pen and moving it around over the yard, the rabbits could be fed with no trouble. Sometimes when we left the pen too long in one position, the rabbits decimated a block of our lawn. Another day when we went out to check on the rabbits, we found three rabbits. Bunny Boy, a rabbit for Ruth, had been slipped into the pen by our grinning father. He was a black and white spotted male rabbit. And you guessed it! We were in the rabbit business. Soon Dad built another wire pen in three tiers, and we had litter after litter of baby rabbits, big ones, middle-sized ones, and little ones. We could never face eating a rabbit, so we sold them and finally turned some loose on the farm. For years we watched black and white Dutch rabbits hop around our pastures.

Perhaps the most formidable pet we had was our dog Patty. It was not her size, but her personality that was to blame—jealous, protective, and excitable. She was a black cocker-spaniel, always ready to play any game we wanted and knew she would make a good house dog. However, Patty was a cat killer. She assumed the role of "Queen of the Yard" when cats came around. No cat could come near her dish. No cat could venture too close to "The Queen" without being shaken. Sometimes we would rescue the cats, but sometimes it was too late. We faced the harsh reality of life. We cried and often, under the direction of Memo (Mattie Heath), held a comforting cat funeral. At times cats would take the place of dolls – with tails. Nancy Puss endured doll clothes and rode around contentedly in our doll buggy. Alice doted on the kittens and carried them around like a mother cat—by the neck. Her small hands would grip them, and as she waggled them across the yard, they learned to be very docile; they had no choice.

The church was a central part of our lives. At first we attended special meetings locally when visiting preachers would come through the area. Meetings were usually held in Cox School. Mother and Dad, who were of different denominations, compromised by choosing the Methodist Church in Lilbourn of which neither was a member. The building was a one-room frame structure with a sloped floor toward the altar. Sunday school classes were held in different corners of the church. A new church

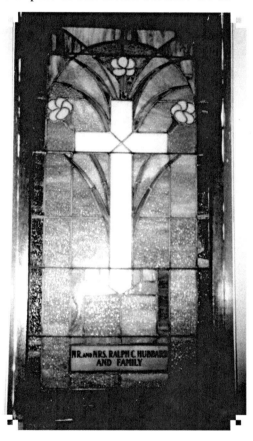

The white frame church built in 1943 was adorned by stained-glass windows purchased as memorials.

was built in the forties with a sanctuary, classrooms, and fellowship hall. It seemed so grand, although it was just a white frame building with lovely stained glass windows.

The Sabbath was an important day in our home. We saved our best clothes to wear on Sunday and made preparations on Saturday night for the next day with hair washing, shoe polishing, and studying our Sunday school lesson. Sunday was a day of rest. Dad would never harvest anything on Sunday. Other farmers would combine crops or plant on Sunday during propitious weather, but not Dad. He would only milk and water the livestock. Women folks cooked, but never did we sew, wash, or iron on Sunday. Those were chores that could wait until the Sabbath was over.

Bible school was a special time. Every afternoon for two weeks in the summer, we gathered up Bible study books, supplies for crafts, refreshments and journeyed to the church for the school. Mother always taught. Other church children in Lilbourn came to our Bible school. The program on the last day featured what we had learned and all the things we had made. The performance was exciting. Mothers were exhausted.

Revival time was another church event. Every summer between the time "crops were laid by" and before "harvest," the church held a revival. It was a time to draw in new converts, a time of fellowship, and a time for renewal. A special minister came to hold a preaching mission. Sometimes the minister was a revivalist and quite a showman—that is, he did nothing but preach revivals. Once a revivalist held a preaching mission in a tent in Lilbourn. He was quite an entertainer—wore a white suit, pranced up and down th[e] painted his fingernails. At other times, area m[inisters] change pulpits. The tone of the revival was one of re[al] sincerity. We would be rejuvenated and ready to walk i[n the] footsteps of Jesus.

When I was eleven, I was baptized. During a revival I responded with others to the altar call to follow Jesus. Because my father came from a Northern Baptist Church and my mother was a member of the Christian Church whose tradition included immersion, my father suggested that there would be no controversy if I were baptized by immersion. I listened to him. Methodists offer the option of sprinkling, pouring, laying on of hands, or immersion. The Lilbourn community was of strong Southern Baptist influence, saying much about the Methodists "who were dry cleaned." So it was that Mary and I joined the group of fifteen people on the banks of Little River one Sunday afternoon to be baptized by immersion. The church congregation sang songs including, "Shall We Gather at the River?" and "On Jordan's Stormy Banks I Stand." Then we waded out into the water for the sacrament of baptism. The minister placed a folded handkerchief over our faces and dipped us under in the name of "The Father, Son, and Holy Ghost." It is a day I won't forget.

The Lilbourn Methodist Church gave our lives focus. Every Sabbath day we were in church. "As for me and my house, we will serve the Lord," reiterated my parents quoting Joshua. "Don't hang your harp on a willow tree," my mother admoned us when Mary and I left for college. Church was where we celebrated every Christmas and Easter. It was where we celebrated weddings and funerals. It was where we heard, "God has been so good to us."

We absorbed the goodness that came through trials and errors and learned to laugh and love in the community called "The Farm."

Chapter 8.
Schools in a New Land

When new frontiers opened, people moved in, and the first question that parents asked was, "How will our children be schooled?" In the growth of every farming community, the building of schools was of prime importance. Plans were set up for school houses to be built at strategic locations along the newly dredged ditches to accommodate children within a comfortable walking distance of a school. No child should walk more than a mile to reach a school. Cox School sat on one corner of the farm adjoining the Hubbard farm, and was the school that we attended. A large bridge across the big ditch made it safe for the children to reach the school. It was named Cox School after Mr. Cox who gave the land. The floating boat called a dredge boat scooped up the mud from the bottom of the swamp and dumped it on a bank, making a levee. The school was built behind this levee.

However, when Alma moved from Wayne County with her parents Joe and Mattie Heath, to the saw-mill community on the Catron ditch road (now Highway W), Cox School had not been built. Alma faced walking to a school located on the intersection of Highway 62 and The Little River Drainage System ditch that ran along the east side of the Hubbard farm. This meant that as an eight-year old, she faced walking one and half miles across the farm on gumbo roads to the school house. All was fine until it started raining and raining and raining. Her heart was broken, for nothing was more exciting than school, yet she could only stand at the window and cry. She had been placed a grade ahead in school, an advance she lost because she couldn't get to the school. Sometimes her dad would carry her through the mud, but other times the ponds were too deep to wade. She stayed at home and worked on worksheets that Mother Mattie contrived out of old paper sacks.

One of the stories that Alma kept coming back to was an account of the last day of school. I can imagine that she had a special dress and was ready for the last day. Spring is a glorious time of the year on a farm. Most of the farm should be dry because the fields were being plowed and planted. Her course across the farm called for some of the usual by-passes, however, and she arrived a little late at the school. All the children had gone home, including her teacher. The clerk in the store that stood by the school house came out to greet her, carrying her report card. He explained that the teacher was so sad to have to leave before she got there, but the bus was there to take him back to Lilbourn. He said, "The teacher wanted you to know you were the best in the class and that he was very proud of you." I am sure it was a lonely trip back across the farm.

No speculation is needed to understand why Joe Heath

would be a member of the Lilbourn School Board. He would help Alma go to school. According to Joe Heath's work ethic, he would have volunteered to build Cox School, if it were necessary. It wasn't long until Cox School was built, and according to the records, Alma Heath is listed in the sixth grade class at Cox School and in the Eighth Grade Exercises on May 13, 1925, at the First Baptist Church in Lilbourn. All the country schools with their eighth grade students were in the school list – Flater, Cox, Woodrow, Mounds, Kewanee, White Oak, and La Forge. Alma Heath and Cecil Hines were the only students from Cox School. Their battle for valedictorian began in Cox School. Alma delivered the valedictorian speech at graduation.

So, high school began for Alma Heath, and the battle with mud continues. We have the story of the school bus that slid in Catron ditch with two wet students, Alma and Cecil. It seems the mud was too slick for rubber wheels to handle. We listened to the story of Joe Heath's ingenuity in rigging up a wagon with benches and transporting students to the bus line with mules. Finally, we learned that two farm families joined forces and rented a two-room bungalow in Lilbourn so their children could participate in school activities all week. Virginia Twitty, her brother Wendell, and Alma honed their home-making skills during the school week and came home on the week-ends. Can you imagine two scholarly seniors and an exuberant freshman living with such freedom? An education was serious business! It meant two years of Latin, Algebra, Geometry, Chemistry, Bookkeeping, Typing, Debate, English, and History, but there was time for fun, candy parties and bunking parties as well. Only the serious students who were college-bound remained in school.

The next lap of Alma's education took place in Cape Girardeau and involved a train ride to get out of the mud and enroll at Southeast Missouri State Teachers College. Mother would entertain us by telling the steps she took to go to college. She lined up a job and took out an insurance policy to cover the expenses she would incur. The job involved baby-sitting for a realtor who worked at odd hours. Shortly before she left for school, she received a letter from the realtor giving her instructions to get in her house. The front window would be left unlocked so that she could lift it and crawl in the window. The mission was accomplished without anyone reporting her to the police. Later in the evening the family came home, and she met the delightful daughter who learned to love the fried-egg sandwiches with lettuce and mayo that Mother fixed after school while they both studied. Alma laughed about seeing her image in a mirror as she walked to school, thinking who is that dour young lady going to class, only to realize it was she. Would she ever get over feeling so apprehensive? Her fierce determination to go to school made her resolute.

Every farm was a busy community. It took many families with children on each farm to do the work by hand. Families moved in to clear the land and grow cotton, corn, and other crops. Cutting down trees and pulling stumps with mules was grueling work for the men. Chopping and picking cotton was tedious work for the women and children. To be task oriented was programmed into every child. To go to school was a special treat. When cotton was ready to be harvested, schools scheduled a vacation to let the children pick cotton. Even little children were expected to pick cotton to put in their parents' sacks. At one time as many as forty children were enrolled in the Cox school that housed all six grades. However, families moved often, and the teaching load would fluctuate to an average of twenty students. Children rode a bus and attended the upper grades in Lilbourn in the adjoining village. Teachers came from Lilbourn on the bus as it picked up older students and rode the bus back to Lilbourn when students returned to their homes. Country teachers were paid more than town teachers because their work days were longer, and they had more responsibilities. It was an awesome task to serve as janitor, nurse, counselor, librarian, music teacher and coach as well as teach.

Marian Pollack was our first teacher. We were fortunate that Mrs. Pollack was willing to take the long, rough bus ride to Cox School. She was an experienced teacher of many years who knew how to organize a classroom of many grades

Marian Pollack taught at Cox School from 1940-1943

and keep all students busy. Story telling was her specialty. She had a fund of stories that she had developed through the years. Every Thanksgiving we listened as she marched across the pages of history and wove Pilgrims, Indians, the British, the Revolutionary War and the out-back life into a marvelous story. It was the days before television when story telling was an art form, and we didn't know that we were digesting history as well. The gallon jar of paste that sat in our "library" was put to use. It was dibbled out on scrap pieces of paper so that we could make a craft to go with the story. What a treasure was housed in our meager library! The tall shelves with doors opened to reveal its treasures—supplies for the school year and a few ragged library books. Holidays were wonderful! We made pictures, paper chains, valentines, Christmas ornaments, Easter eggs and May baskets to go with the stories she told. We became a part of history and walked a little taller as we grew in understanding our world.

Perhaps you would not call Mrs. Pollack glamorous, but when we played school, we thought acting like Mrs. Pollack made us perfect teachers. She was tall and thin with a nose that matched her stature. Her puffy white hair softened the bony outlines of her face, and the white handkerchief she swished around her nose made us think that long thin noses were most attractive. Always she carried one with lace on one corner. Handkerchiefs seemed very important to her. Of course those

Marian Pollack was a master at organizing a one-room school.

were the days before Kleenex. We tried to remember to bring our handkerchiefs, or snot rags as students called them, but always we forgot on the day we needed them most. Our only option was to sniff. Mrs. Pollock really had an aversion to sniffing. I remember the day she stopped teaching and gave the whole class a lecture and demonstration on how to blow noses. We got out our handkerchiefs or rags and practiced holding one nostril, blowing it, and swabbing out the debris, then holding the other nostril and doing the same. Can you believe that I still think of Mrs. Pollock when I blow my nose? We wanted to please her.

A paddle always hung in a prominent place in every country school, especially the one-room country schools. However, I cannot remember Mrs. Pollock ever using it. She used her eyes. Always her deep brown eyes roamed over the classroom while she listened to classes recite in the reading circle, or when she checked work on the blackboard, or when she gave out spelling words. Her eyes made time-outs unnecessary. In some schools, teachers made students sit in a corner with a dunce hat, or so I've been told. But that wasn't Mrs. Pollock's style. When her roaming eyes stopped on a student like a dog on point, we all got quiet. If a student didn't notice, she would bite her lower lip, tip her head down, look at the student over her glasses, never moving her eyes a blink. When the student caught a glimpse of her eyes, we could count on watching the culprit slide down in his seat and pretend he did not know Mrs. Pollock had seen him. Sometimes she would bite her lip and shake her head. That was bad news. When her eyes looked at me, you can be sure I started to work. I wasn't going to have the whole class looking at me.

Six grades were taught in one room. If we got bored, we could listen to what the other grades were doing. When Mrs. Pollock talked about days of exploration, she made us want to join the crews and paddle up the river. If we had trouble figuring out a math problem, one of the older students would lean across the aisle and help. By the time I was in third grade, I became a "leaner." I would lean over to help a second grader pronounce a word or work a math problem. The process of teaching began at an early age, a process called "leaning." Every grade had its own row of seats. Every year each class would move over to the next row of larger desks.

Life was not easy in the early days that land was being cleared for farming. Families struggled just to have food and clothing. School was a retreat from the work world. At school we could play with the children who lived on our farm and the surrounding farms. It was worth going to school just to have supplies and recess—time to play new games, to read stories, to sing, to cut with scissors, and to color with the new crayons.

The schoolyard was an adventurous place. The shiny, red pump under a big cypress tree was always great fun. At recess we took turns pumping water for each other with the big pump handle. On hot days, we'd stick our faces under the stream of cold water and let the water splash on our faces as we gulped at its stream. It was our manual drinking fountain. On extra hot days, we'd dip our bare feet in the puddle under the spout, or take sticks and dredge a ditch for the run-off. If we had time

Barefoot days came early in the spring at Cox School.

we might take clods of gumbo and build a dam. Always our recesses were stretched to accommodate our creativity or adjusted to meet the demands of the weather.

In the winter or on rainy days, we would pump a bucket of water to carry into the school house. The bucket with a metal dipper sat in the back of the room on a shelf with an assortment of glasses and jelly jars. It became our drinking fountain. We got drinks by serving ourselves from the bucket with our own glasses. At noon we had glasses of water with our lunch. I remember feeling sorry for the children who had sandwiches made with peanut butter and cornbread. I did not like cornbread or peanut butter and was glad my mother could make biscuits. We learned compassion at an early age.

At the back of the lot were two white outhouses to match the white schoolhouse. A privacy fence wrapped around them, one for the girls on the north and one for the boys on the south. What a place to be companionable! On the "three-holers," we talked about all the things we wanted to do. The world was beaconing us.

Directly behind the school building was the coal shed, a most wonderful building of just the right size to play "Anti-Over." The roof sloped ever so gradually but was tall enough to be challenging to the "throwers." Teams took their places on opposite sides of the shed. One side would call, "Anti," and the other would respond with, "Over," signaling a readiness to catch the ball. The game was on. A thrower would heave the ball over the roof. If someone could catch the ball, the whole side would run around the shed and try to capture players by hitting them with the ball before the team switched sides and were in safe territory. And so the game continued.

The coal shed holds another story and my first introduction to "The power of being a woman." Most farm families burned wood because wood was plentiful on the farms, and the smoke was much cleaner. Cox School used coal because it made a hotter heat and would burn longer than wood in the big potbellied stove with a metal jacket. Every school day during the winter, the farm hand who lived close to the school built a fire so that the building would be warm when the school day started. The rest of the day, the teacher would see that coal was added. Behind Cox School the small coal shed was built to match the white-framed school except it had no windows. Windows in the school building were built only on one side of the structures, near the ceiling so that children would not be distracted by outside activities while they worked. Schools were designed to let the light in, but only the teacher could look out.

It was the custom for the older boys to go and get coal in coal buckets. That meant getting out of class, putting on coats, shoveling the coal in the buckets, and bringing them into the classroom. One day when the teacher asked two boys to get more coal, Laura rose, stood by her seat and exploded, "Why can't girls go get coal? I would like to get the coal. Why do just the boys get to go?" We all stopped and looked at Laura. The class was silent and thoughtful. Then we looked at the teacher. She paused, and then said, "I see no reason why the girls can't get coal." Mission accomplished. I shoveled the coal bucket full and helped carry it into the building. It was hard work, but I felt a smug pride in doing so. The feminist movement was launched.

My favorite school story is about the Jacksons. The Jacksons lived in the tenant house behind our home, next to the farm shop and mule barn. Mr. Jackson was my father's most capable farm hand, a shy, quiet man, good with livestock and equally capable when it came to planting or harvesting crops. The Jackson kids were super little cotton pickers and made a game of everything. I really didn't know Mr. Jackson very well, for my interest was in his children. He had eight of them. My sister and I could hear their laughter across the pasture and watch them run in and out of the porches on their white, four-room house. Every year our mother cat had kittens, and Mrs. Jackson had another baby. While my sisters and I played with the kittens, the Jackson children played with a new baby. In pleasant weather when the cows were not in the pasture, we would join the Jackson children in the neutral territory of the pasture to play between our houses. They were experts at climbing trees, at hiding, at kicking cans, and plotting new ways to play tag.

Sometimes Junior, the oldest boy of the family, would fashion a toy out of the scraps around the farm shop. He nailed a tin can to a board and pushed a hoop with it. We, too, rounded up tin cans and spent hours running barefoot while rolling a hoop from a wagon wheel down the smooth dirt roads. Junior made stilts, and we learned to walk on stilts. Our feet soon developed the same hard pads that the Jackson children had, for they never had to wear shoes as soon as spring came. Their mother never seemed to worry about colds. They never had to worry about viruses or allergies because they didn't know about them.

Often we would walk to school with the Jackson children. Sometimes we would stride purposefully down the gravel road to Cox School, across the wooden bridge, and into the building. But during the dry season, we would hike down the middle of the drainage ditch and explore the deep valley cut in the swamp that was dry in the summer and boasted a profusion of wild flowers. Mrs. Pollack made us think every fistful of violets was beautiful when stuffed in a jelly glass. That's when I first noticed Betty Jackson's dress. It had no buttons, no zipper, no sash, or no draw string. She was sewed in her dress. I looked at the other Jacksons. Patsy's dress had buttons. Mae's had a zipper. The Jackson boys had most of the buttons on their shirts, but Betty, the peppy little first grader, was sewed in her dress. I said nothing, for never did we discuss clothes. Clothes cost money. There was never enough money to keep up with the need for new shoes, bigger shirts, longer trousers, warm coats, and socks without holes. Families made do with cast-offs, and no one commented.

At school we were too busy playing to worry about what someone did or did not have on. At noon we gulped down our lunches, often wiped our hands on our clothes, and ran out to play on the acre surrounding the school building with the coal house and two outhouses. The older students watched out for the little ones, most of the time, just as they did at home, most of the time. Little ones were involved in games as the need arose. I remember being recruited to play softball because the fifth and sixth graders needed another person for a team. I had never hit a ball with a bat. In fact, I had never held a bat. I was from a family of girls and was more interested in making mud pies under the oak tree. Some worldly-wise sixth grader who had learned the art of sweet talk convinced me I was big enough to play softball. I dutifully stood while she helped me hit the ball and then dutifully ran to the paint-can lids that served as bases while the team cheered. I liked the cheering and decided when I got so I could hit the ball with a bat, I would like this game with the paint-can-lid bases.

When I got home, my mother greeted me at the door with the usual question, "Well, how was school today?" I was more interested in looking at the new kittens, but I reported, "I played softball with paint-can lids, and Betty Jackson was sewed in her dress."

The next day we were almost half way to school when the Jacksons ran to catch up with us. And there was Betty in the same dress sewed on her. I couldn't help but ask ever so discretely, "How do you get out of your dress when it is sewed on you?"

"I don't," she said in a perky voice.

"But, what did you sleep in?" I asked.

"My dress," Betty answered.

I looked at the dress. It looked clean, well mostly. Then I watched as my little sister Ruth walked all the way around her friend Betty, looking ever so pleased. We had another busy day at school with recesses that were much too short, but we dutifully came in when the teacher rang the bell. I didn't have to play softball because the sixth graders organized everyone into playing dare base. We stood in two lines facing each other and dared the opposing side to come off their base into the middle territory where we could tag and capture recruits for our team. This was probably our favorite game. The whole school played with the daring fifth and sixth graders capturing the most prisoners and setting free the most captives. The first and second graders were willing captives, pleased by the attention of being caught and changing bases.

After school when we banged the screen door at home, Mother appeared quickly. Ruth did the reporting, "Guess what, Mama, Betty Jackson's mother sewed a dress on her. She doesn't have to take it off at night. She just sleeps in it."

My mother laughed, "Yes, I know. Mrs. Jackson brought her new baby by to show me and told me about sewing Betty in a dress. She couldn't find any buttons and someone had worn all the clean dresses, so she just sewed the dress on Betty. At least she sent her to school."

"Oh, mother," my little sister pleaded, "will you sew me up in a dress tomorrow?"

"Only if you promise to go to school and keep it clean all week," my mother replied with a twinkle in her eye.

Soon the Jacksons with their many children vacated the four room house, and the Yarbros moved into the house behind ours. What fun to have children our age to play with. Demetra and Joan Yarbro became our close companions. They loved books as much as we did. Joan and Ruth entertained themselves in the cotton patch by telling stories while they picked cotton. Demetra joined Mary and me in going places together.

And so we learned socialization in the gentleness of a woman who taught with compassion, and in the give-and-take of the Jackson and Yarbro families who shared their cheerful acceptance of life and its exciting possibilities. Ours was an educational system that grew out of watching adults struggle to solve problems. We, too, learned to solve problems and to ask for help.

School buildings serve the community in other ways. Religious groups took advantage of the vacancy of the school buildings on Sunday and moved into each farming community to reach the churched and the un-churched. Preachers were delighted to find a ready-made congregation willing to listen to them, and so churches were born.

Would-be politicians used the school houses as forums for reaching people and fostering their ideas in the world. The Agriculture Extension Service moved into farm homes teaching homemaking skills and beautification and sharing the latest research on farming. Thus aesthetics were born in the rough hewed buildings, and we learned that life was good

Chapter 9.
Stretching for Space in the Old Farm Home

The original farm house was built on a tract of land purchased by Fred and Martha Hubbard of Urbana, Illinois, three miles south of Catron, Missouri, on Highway W. It was a two bedroom home built for the caretaker of the farm. The Richard Trimble family of Urbana, Illinois, moved into the home in 1918. After an unsuccessful venture by the Trimbles of attempting to tame the new land, Joe Heath agreed to take over management of the farm and moved into the home. Joe's work ethic was what was needed on the swamp land. When he was twelve years old, Joe had gone to work in the sawmills of Wayne County and held down a man's job. Each week he would deliver his check to his mother who would save a dime of every dollar for a rainy day and use the rest for family expenses. He knew the value of hard work. When Ralph and Alma married, they lived in the fawn and brown house with the shake roof, and Joe and Mattie Heath moved to a home built near the mule barn. Mother use to laugh and say, "I've lived in the same house since I was twelve, but I didn't get married when I was twelve."

The original farm house sat behind the big ditch.

You can see this house inundated by flood water from the deep ditch in front of the house with water standing in the yard. This is what happened after every sizeable rain. We waited a few hours until the water in the ditch fell, and the yard and fields drained. A screened porch is to the left of the front door, and a living room is to the right. As a child, I remember swinging in the big swing that hung on the end of the porch and watching the sunset. I liked swinging but could do without the running commentary on the beauty of the sunset. Once when our Hubbard cousins visited us, one exclaimed, "Oh, look at the sunset! Look at the sunset." My sister Ruth dismissed her with obvious boredom, "It does that every night." In the flat country the whole panorama of the sun dropping below the horizon banked in clouds offers a display that hill people miss when their sun drops behind a hill. It was like watching a fireworks display evening after evening.

Bedrooms opened off the kitchen and dining room so that heat could be siphoned into the rooms during the cold weather. The dining room was heated with a wood-burning stove, and a wood range was a part of the kitchen heating and cooking system. A warming shelf above the range kept foods warm, and a reservoir in the stove kept a supply of hot water. The water system consisted of a kitchen sink with two pumps—a pump to the tank of rainwater under the house and a pump driven into the ground as a source of drinking water. Water was carried from the pumps and stove into the bathroom for baths. A chamber pot took the place of the toilet. Dad remarked that the Hubbard boys thought going to the farm was equal to going on a camping trip.

At the back of the house was a porch and a junk-room. The gas refrigerator was housed on the porch—a real luxury in the pre-electricity days. I remember the reverence in the tone of voice that mother and dad used when talking about "The Refrigerator," a mysterious affair that I was not to open. A swinging stairs came down from the attic to the porch. Interesting things were stashed away in the attic. At times Mary and I climbed the stairs to watch Dad prepare boxes for his beehives. He bought sheets of man-made honeycomb, then cut it to fit wooden frames for the beehives. We loved to watch him weld the sheets into place with candle wax to prepare the frames to fit into the beehives. While there, we enjoyed prowling through the junk in the attic, especially the boxes of Christmas ornaments. The porch was where Mother rolled out the washing machine, set up tubs, and washed clothes. It was an all day operation that we delighted in supervising, if we didn't have to go to grandmother's house. The junk-room always provided the most interesting things for a child to prowl through. I don't remember when the bed came, but suddenly there was a day bed in the junk-room. Mary and I were informed that Cora was going to live with us because mother needed help with the work. What we didn't know was that the work involved taking care of a new baby—Ruth.

In the back yard, a functional affair of wooden steps and a wooden platform caught the mud and grime of the farm. All gumbo boots and work clothes were deposited on the platform before they were brought into the house. The garage was a lone wooden structure in the front yard that housed our one car and sat on the ditch bank beside the wooden bridge across Catron ditch. I can remember waiting beside the car while Dad cranked it up to get it going. It was hard work starting a car with a crank, no automatic switches. We learned patience.

The dining room served as a family room with a fold-up table. Doors from the porch went both into the dining room and living room, for circulation of air was important in the days before air conditioning. As long as I can remember, we had a piano in the living room. Mother would play hymns, and Dad would play his clarinet, transposing the music to play in the right key, most of the time.

In 1942, Mattie and Joe Heath built a house on Highway W across the pasture from the Hubbard home. I remember their moving day. Mary and I made some of the trips from the old

The Heaths' new home was built in 1942.

A three prong cypress tree made an ideal fort.

house to the new. Ruth, who was three, hunkered down in the middle of the flat-bed wagon and was so pleased to ride with the furniture to the new house. Quickly we had a path worn in the pasture grass between the two houses, through the chicken yard and over a foot bridge. We had a new area to explore, a new attic in which to build a play house, a new garage with a smokehouse, and a complex of three cypress trees that had grown together with knobby knees for a fort.

The house behind the cypress tree was remodeled to have two front doors.

Always a remodeling job of some kind on the house was in the works. The front porch was enclosed, turning it into a living room in order to resurrect another bedroom. The dining room and living room were lengthened. Dormer windows were added to the front room and the attic. A front door was placed on the corner of the house, and a stair was built to the attic in the old front door area. The attic was finished off. Closets were added to bedrooms. Hardwood floors were installed. The fawn house was covered with white enameled siding, and the old shake roof was replaced with a synthetic one. A new bath was added as well as another back porch. Finally a two car garage was built behind the house.

In 1961, after Joe Heath's death, Mattie Heath built an apartment connected to the Hubbard Home with a porch. It consisted of a kitchen, bedroom, bath and a combination living and dining area. The apartment is to the left of the main house. Both of the two former homes of Joe and Mattie Heath burned.

This sketch of the farm house where Alma, Ralph and their four daughters lived for over fifty years was drawn by Elizabeth Russer of Puxico, Missouri.

In 1983, after Ralph Hubbard's death, a garage with an automatic door opener was attached to the house. The final stretching had been accomplished.

A side view of the farm home shows an attached garage.

Four generations lived in the Hubbard farm home–Mark, Mary, Alma and Mattie.

Chapter 10.
Gone—Barns, Fences, Gates and Stiles

In China we have the Great Wall because people are obsessed with boundaries. On a farm we have our boundaries marked by fences and gates, dividing the farm into work units. Fences define yards for houses, gardens, and pastures. Fences kept hogs, cattle, and chickens from wandering. Fences marked the boundaries of fields and ditches, and fences kept children safe from the roiling water in ditches.

As a child I grew up within the fence that surrounded our house, marking the yard in which my twin sister Mary and I could play. I was fenced out of the garden and orchard. I was fenced off the big bridge and the ditch that ran in front of our house. I was fenced out of the chicken yard, the corn crib, the shop, the mule barn, and away from the tools and wagons. Behind fences we could watch the bustle of the farm hands loading wagons, hooking up mules to tools, starting tractors, scraping off mud, and making repairs. We were safe. Fences gave us boundaries.

Gates offered a way to move about between bound-

Posts, gates, and a fence set off the boundary of the garden from the farm house.

205

aries. Gates were heavy wooden affairs that would stop an aggressive bull and that only a strong adult could lift and swing open. Sometimes men would save time by climbing over the gates rather than taking time to unhook them and swing them free. Joe Heath was amused at Ralph's youthfulness in vaulting over a gate. An interesting contraption in the middle of the shop area was a stile. It was a way of walking over a fence. Its efficiency pleased me, for men could walk up the huge steps, over the fence and come back quickly without opening a gate. However, as a little girl, the high steps of the stile were daunting to me. Sometimes Mary and I would sit on the stile and feel safe while we watched the mules and the workers. Once, Dad put us in a wagon to keep us out of harm's way. Some cows sauntered by, and one "mooed" in our faces. We were too frightened to scream and so relieved when the cow lumbered away. Dad saved us!

Fence rows were a wonderful part of the farm. As we grew older, we explored the fence rows, looking for wild flowers growing on the fences. We marveled at the morning glories that were triggered to bloom by the sun. We collected milkweed pods and turned the fluff of the seeds into fanciful, wind-blown creations. We chased rabbits out of the brush and drank from the pump driven in the fence row to satisfy the thirst of workers in the fields. We thrilled at finding blackberries growing in the weeds and on the fences. We inspected the Catalpa trees that had been planted in the fence rows to augment the fence posts. This practice, touted by the Extension Service, was wearisome as the trees grew larger than expected, and the chore to take out the trees was tedious when the farm changed from a stock farm to a rice farm. Fortunately, Dad had not succumbed to the suggestion that he should plant multi-floral rose as a fence and bird cover. He anticipated the prolific growth of the multi-floral rose planted in the fertile swamp soil and said, "I don't believe I'll do that." It was in the fence rows that we picked asparagus that was seeded by the birds from our garden plot, and it was in the fence rows that the pheasants hid when Dad turned them out of the chicken yard. Native instinct kicked in, and the pheasants found safety and protection in the thicket of the fences. Fence rows were a source of endless treasures.

One day when I was twelve, Dad announced, "How would you like to drive the truck to 'The Center'?" I caught my breath and gulped. Was I really old enough to drive? I had reached my height early, but oh, my! "The Center" was the magical place to which Dad went everyday, sometimes several times during the day. It was the place that Molly Brown, the horse, took Dad through the mud when it rained. It housed three barns, two tenant houses, but most important, it was central station for the two livestock tanks. In the flat land that had no stock ponds, watering tanks for the cattle were a prime necessity. "The Center" was where the cattle were fed. Going to "The Center" of the farm would mean opening and closing gates, a chore that would be helpful to Dad if I went along. It would mean traveling down dirt roads that wound through fields, staying in the ruts, circling around trees, and getting through gates without knocking down any posts. Oh, I was ready!

Mary is ready for her driving lesson in a battered farm truck.

The first trip was in the battered pick-up truck. I listened to a lecture on shifting gears, using the brake and accelerator, and we were off. We eased through the gate by the house, creeping along at an exciting two miles an hour to the second gate by the farm shop, and we were on our way. I even shifted into second gear. The road was gumbo, packed hard by wagons and trucks and molded into ruts during wet spells. We dutifully started the pumps which would automatically cut off when the tank was filled. Eagerly I got back in the truck and wended my way back through the gates. I would soon graduate to the bigger stock truck that called for more skill in getting through gates without wiping out a gate post. So I had learned to negotiate my first gate and was ready for the gates that would open to me throughout life, some heavy and some light wire-affairs that would swing open to reveal roads I would travel in the years ahead. I stood tall enough to take my place beside the adults.

Barns are another important part of the landscape of any farm. Near the farm house is an assemblage of the important work areas for the farm—the two mule barns; the big red crib that housed the corn crop; the blacksmith's shop where equipment was repaired; a tool shed that housed the tractors, combines, hay-baler, and other tools that needed repairs during the winter; and finally the seed house where seeds were cleaned, measured, and sorted. The buildings were wonderful places to

Heaths' new barn is in the upper left hand corner of this aerial view of the farm.

play hide-and-seek on a Sunday afternoon when children came for Sunday dinner and stayed to play on the farm. There were untold places to hide among the tractors, the tools, and the spider-webbed corners of the barns.

The newest barn on the farm was built on Joe Heath's farm to augment the limited space in his small barn. When the new barn was finished, someone suggested a barn-warming. So, on a Sunday afternoon, the people on the farm brought their fried chicken and side dishes to the new barn. We ate at improvised tables built out of left-over boards. Afterwards the children bounced and slid on the fresh hay bales. Well do I remember when we decided it would be fun to sleep in the barn while it was new and before the spiders moved into new hiding places. So it was that we brought our blankets, pillows, and flashlights to spend the night in the barn, a real adventure, and the only time I ever slept in a barn. I felt like a real pioneer.

The feeding-lot barn on the northwest side of the farm housed the steers. I see in my mind's eye the line-up of cattle munching

The barn for the steers was dismantled after Ralph closed out his cattle operation.

away at alfalfa hay and "tankage," smelly stuff that was supposed to make cattle gain weight. Little did I realize what gaining weight would do for them. I remember Dad commenting that it was a "blue" day when he sent his cattle off to market. He never wanted to butcher an animal and said he would buy his hamburger in a store. The "mothering" barns were at the center of the farm, three barns where cows would care for their calves. So the parade of cows was marched around the farm to various barns and to various fields. The corn fields were cleaned up of nubbins, husks, and stalks, and dutifully the cows served as fertilizer spreaders. The hogs were housed in A-frame houses that could be pulled around to clean ground to keep the animals free of disease. Livestock was always a part of Ralph's diversified farm operation. The land of the lowest fields often served as pasture when arable crops drowned. Holes became mud lollies for the hogs. Ralph commented, "Raising cattle makes farming fun." As he grew older, he quit raising hogs in 1955 and slowly phased out his cattle operation.

An early history of the farm is documented on the application form for the "Plant to Prosper Contest" sponsored by *The Commercial Appeal* of Memphis, Tennessee, in 1938. Alma Hubbard writes:

This farm, an un-drained timbered swamp, was bought by Fred Hubbard (Dairy and fruit farmer of Urbana, Illinois) in 1908 for $18 per acre. It was drained and cleared, and until 1931 was under the supervision of a manager (Richard Trimble). During that period the farm never paid. It was estimated that it had cost $100 per acre by 1931, and there was a $35 per acre mortgage on it. This had made it necessary to also mortgage Fred Hubbard's Illinois farms. As the depression came on, it seemed the Missouri farm would go to the mortgage holder. To relieve his father of this burden, Ralph Hubbard came to Missouri (1930) and took over the management of this farm. To save his Illinois farms in case of foreclosure and because it seemed the Missouri farm was lost, the father deeded it to the son (Ralph) to do with it as he chose. In addition to the mortgage, there was an $1800 crop loan, a manager's salary due, delinquent drainage tax for two years, county tax for one, and a delinquent loan payment of $600. The machinery inventory was low, and the buildings in need of repair. Ralph came to the farm to live with no asset except what the farm produced. Now, 1938, all loans are paid up to date, taxes are paid, machinery inventory increased, buildings in much better repair, livestock *inventory doubled, and just this year the loan was refinanced at 5% interest rate instead of 6%.*

The farm was first called the Flagland Ranch. When Ralph Hubbard started farming, he included livestock. This was reflected in the label on the farm stationary:

Flagland Stock Farm
Registered Duroc Jersey Hogs, Direct Line from Stilts Wavemaster
Hereford Cattle

Additional acreage was added to the original purchase to make a total of 949.44 acres. After the debt on the farm was paid off, the farm stationary reflected another definition:

HUBBARD STOCK FARM
RALPH C. HUBBARD
ROUTE 1 - BOX 100 - CATRON, MO.

Hubbard Stock Farm
Registered Polled Hereford Cattle
Registered "Easy Feeding Type" Duroc Hogs
Lespedeza Seed

The mortgage on the home farm had been paid off in 1943, so the time was ripe to add more land. Ralph Hubbard said he would like to see what he could produce on sandy land. The experiment began. In 1944 Ralph and Alma Hubbard bought a farm near Risco of 160 acres and named it Shady Grove. (A tract along Highway 62 of this farm was later sold for a subdivision.)

All of Shady Grove was arable. Corn, cotton, and soybeans became the trial crops. When the years were wet, this land produced well. When the season tended to be dry, the heavy gumbo farm out-produced the sandy soil. It was not long before Ralph reached the conclusion that Shady Grove would be suitable for truck gardening, if it were put to grade and irrigated. The process of grading the farm started. Half of Shady Grove was graded, not a really good job, for farmers were just beginning to tackle putting land to grade. (Grading was the process of moving the top soil of a field so that the crown or highest part of a field would slope downward into ditches and drain uniformly.)

Ralph and Alma were right in concluding that to grade a farm took a lot of the risk out of farming. To be able to water the crops in the dry season made all the difference in producing a successful crop. To drain a field during the wet season saved planted crops. Picking cotton on sandy soil was comforting to those of us who had endured the hard clods of gumbo like rocks on our knees. Soon, however, the mechanical cotton picker moved in to take over the hand picking. Ralph lamented the fact that much cotton was wasted in the process of mechanical picking. New varieties of cotton, however, are adapted today for the mechanical picker.

The Shady Grove farm had evidently been part of an Indian campground. The Baker boys, who lived in the tenant house, collected arrow heads each year after a new crop was cultivated.

After Joe and Mattie Heath died, their farm of 152 acres was added to the Hubbard farm. (Joseph M. Heath died in 1959, and Mattie E. Heath died in 1978.) They had purchased their farm on February 16, 1934, during the depression, for $8 an acre. At one point the Heaths swapped some of their land for Hubbard land on Highway W, so they could build a home on the Catron ditch road (Highway W) and be near the Hubbard farm house. They build a white frame home with two bedrooms and an upstairs. This house was burned by a renter after Mattie Heath moved out of it into the apartment attached to the Hubbard family home. Today only the remains of a wooden bridge across the ditch and cypress and oak trees mark the spot where the house once stood.

The land was incorporated in 1983 after Ralph Hubbard's death and set up as a Sub Chapter "S" corporation named Heath-Hubbard Farms, Inc. Part of the farmland was transferred into the corporation in 1983 after Ralph's death and the remainder was transferred into the corporation in 1990 after Alma's death. The land is figured now as:

Ralph Hubbard	278 Acres
Ralph & Alma Hubbard	144 Acres
Alma Hubbard	514 Acres
Shady Grove	157 Acres
Joseph & Mattie Heath	<u>151 Acres</u>
Total	1,244 Acres

Land grading became the primary concern in the 80's and 90's, and the face of the farm was changed. The farm was taken out of the government program to grade the fields and build a crop base for rice. On the Catron farm, fences, barns, and houses were removed, trees taken down, ditches dredged and a road was set up to go through the middle of the farm. On the Risco farm, the other half of Shady Grove was put to grade. A schedule to grade all fields was put into action. The following renters have helped to make this possible: Bill and Kenny Denton; Dick Twitty; Steve Lancaster; Andy and Rick Smeltzer; Billy and Bill Chamberlain; Don

Stubble in the fences was cleared when the farm was put to grade.

Westmoreland; John and Richard Burnett; and E.P. Priggel and sons Jim and Mike.

The farm was no longer the vibrant community of families in tenant houses with barns and animals that I experienced as a child. The fences were rolled up in big bundles and taken off to be recycled. Old implements stashed away under shade trees were sold. No longer were they needed for used parts. The cattle tanks were buried. The junk pile where we hunted for treasures among the junk dumped by the tenants through the years was buried, and the land was put to use.

After Alma's death in 1990, my husband, Bill, a C.P.A., took on the task of serving as farm manager, learning all he could about farming from his farmer clients. We spent much time at the farm. The fact that the University of Missouri had a research farm just eight miles south of the family farm made collecting data for decisions convenient. By putting the farm to grade, our drainage problem was solved, and irrigation was put in place. We found that farmers in the surrounding area were tackling the same problems we were addressing. Rice is a popular crop in Southeast Missouri because the high water table makes irrigation simple and inexpensive. A rice buyer on the Mississippi River makes marketing the grain convenient. However, we learned that rice farming is work intensive. Checking wells and water levels is an exacting task. So many decisions were new to us.

We set about learning how to raise and harvest rice. I love to see the serpentine levees winding their way through a field of rice, but learn that straight levees are more efficient because they indicate that a good job of land grading has been accomplished. As I watch the green shoots of rice fill out with heads that bowed down under the weight of the grain and undulate in the wind, I speculate on how much the rice might yield per acre. Planting levee rice on the top of the levees helps me mark the various pads of rice. When I see rice lodging (falling down), I agonize about how it will be combined. "Tediously and with some loss," I am told.

As a well is driven, I watch the process of providing water to the fields. A drilling rig moves in. A small pipe is driven down about forty feet to get water to use as a lubricant, while a big well of sixteen inches in diameter is drilled to a depth of a hundred plus feet. Everywhere our farmers came up with answers to rice problems, and rice specialists supplied the research data.

I wish that Ralph could see the harvest of the big graded fields in which he had so laboriously pulled weeds. Fall is a beautiful time of the year. The patchwork-quilt effect of contrasts between various crops in the fields are laced together with roads. The green rice fields turn gold; the cotton bursts from its bolls; the soybeans turn brown; and the corn bulges in its dried shucks. Ralph would have been thrilled by what controlling the water on the farm has accomplished, for he and Joe Heath worked so hard after each rain, ditching the water holes to save the crop and get a better yield. At one time he planted what we called "the forest" to provide shade for his hogs. It was a half acre of trees in the lowest land on the farm, the part Joe Heath called the old "Castor River" bed, not worth planting. The pigs had a

Beth finds climbing on a tractor as much fun as a jungle gym.

marvelous time turning over the forest mud with their snouts and wallowing in the mud.

Going through a sea of rice on a lumbering combine makes me think I know what the astronauts experienced when they tried out the electronic moon buggy. We are always delighted when we find our renters, the Priggels or Andy Smelzer harvesting a crop. We get ready for a show-and-tell session. It beats a Ferris wheel ride any day. Once when we found the Priggels in a rice field, we climbed aboard their massive combine equipped with a stripper header to make the rounds in a field of rice. Half of the rice was Cocodrie and half was Cypress. Priggels explained that the Cocodrie is a short rice that does not lodge (fall down)

Watching a stripper-header comb out the rice and leave the stubble in fields to enhance the soil makes a farmer cheer for technology.

and matures three weeks earlier than Cypress rice. Having two varieties of rice helps their harvest plan. The Cypress rice is a tall grain variety that was still quite green. Mike Priggel drove the combine and touted the efficiency of the stripper header that is in common use on rice farms. He explained that the harvest is much faster with the stripper-header because the energy of cutting the stalk is bypassed. The heavy, draping heads of grain are combed up and suctioned into the combine bin in a vacuum-cleaner action. The rice stalks are not cut but remain in the field, making one think on first glance that the field has not been harvested. The thick rice stubble is considered a boon to improving the soil, especially when the ground is rolled to help the stalks rot. Ralph would approve. The combine with dual seats in its cab, is air-conditioned, has a panel of electronics that makes me think a lift-off might be possible. We sit up so high we can see for miles – even better than the view from a mule.

Jim Priggel supervises the unloading operation. The grain from the combine is dumped into a holding basin and an auger is used to lift it up and funnel it into a grain bin. It makes me recall the days I stood as a child and watched the corn go up an elevator into the big crib by the house while the mules powered it by walking around a pivot. Now, everything is done with electrical or diesel engines.

When our daughter Teresa brought her children, Luke, Ellie, and Zora, to the farm, they picked cotton on Shady Grove to take to school for show and tell, and ate some of the corn in the roasting-ear stage, right in the field. What a tantalizing treat!

Today's farm is one of fine tuning possibilities. Ralph dug ditches, moved dirt for roads, tried out new seeds and worked at improving the soil through diversity in farming practices. Our renters, E.P, Jim and Mike Priggel and Andy Smelzer clean out ditches, shape up farm roads, try out new varieties of seeds, manage irrigated fields, and try out various crop rotations. The battle that started out to control swamp water, now focuses on controlling the water of irrigation.

Gone are the fences and gates and stiles. My heart catches in my throat when I drive through the farm and realize it looks nothing like the bustling farm I grew up on. I see vast fields that have been burned and plowed to get rid of the red rice and sheath blight, following the advice of the agriculture specialists. Bill consoles me with, "But it is productive." Gone are all the tenant houses, the barns, the trees, the fences, the chickens, the gardens, the pumpkins in the corn field, the wagons, tractors, and hay baler. Now I see huge combines, cotton pick-

Gone are the barns, replaced by grain bins.

ers, eighteen-wheel trucks, twelve-row planters, and a massive roller.

I hear my grandson Justin's voice ring out when I tell him that he can run on the farm and explore anything he wishes, "Oh, Neanie, you don't know how lucky you are to have lived on a farm."

Yes, I do, Justin. The farm was a laboratory that shaped my life. It gave me walking space in the dirt, in the water and in the mud. It grew hollyhocks and corn shucks to make dolls. It hosted a place we could find caterpillars, walking sticks, crawdads, dragonflies, and grasshoppers that chewed tobacco. Ditches of all sizes became subterranean pathways through fields to be explored. We crawled through culverts and chased down rows of corn. We went on treasure hunts in the barns, in the seed house, in the blacksmith shop, and in the tool sheds. We climbed trees in the orchard and tramped down cotton in a wagon. We rode the combine and helped catch soybeans in a sack. We punched wires on the hay baler. Animals entertained us with their antics – kittens, rabbits, pigs, chickens, ponies, calves, mules, and baby "gobbies" (turkeys). Always there was the big garden to traipse through and find a tidbit to taste.

Gone are the fences and gates and stiles. Gone are the tedious tasks associated with a farm. Ahead lays the technology that gives the farmer the freedom to produce an abundance of products in a modern environment and wrestle with the figures of production and cost. Let us treasure what the farm families of the past have given us.

Yes, Justin, I know how lucky I am.

Epilogue
by William D. Stewart

Ralph and Alma (Heath) Hubbard had four daughters: Mary Ellen, Martha Jean, Alma Ruth, and Marjorie Alice. Mary and Martha are twins born on February 12, and Ruth and Alice were born six years apart on June 16. This meant the family had four daughters with only two birthdays. It was my good fortune to marry Martha Jean on one of the June birthdays. Through the years I had many conversations about estate and financial planning with my mother-in-law, Alma E. Hubbard. As a CPA, I had an accounting practice in Poplar Bluff, Missouri, and worked with many farmers. Farmers are special. They work long, hard hours, take enormous risks, and their efforts are always subject to the whims of nature. Yet, they persist and are always willing to help others. As business manager of the family farm, farmers were my best source of information and suggestions to improve the farm.

Mother Alma always made certain things very clear. She wanted her daughters to benefit equally from the family farm, and she wanted the farm to remain intact, including the land of her parents, Joe and Mattie Heath. These were the two problems she didn't know how to solve: that all the soil is not equal in quality and that dividing the land would ultimately result in dividing and breaking up the farm. Often, difficult problems have simple solutions. Such was the case. After Ralph died on July 6, 1983, I explained the possibility of having a family corporation own the land and distributing stock ownership equally between her four daughters.

She saw the formation of a corporation as the perfect solution to her dilemma, so on October 11, 1983, a family corporation was formed. Alma chose the name "Heath-Hubbard Farms, Inc." to honor the families who made it all possible. After Alma died on January 25, 1990, the land was transferred to the corporation, stock was issued one share per acre, and the daughters each owned a one-fourth interest in the corporation. The certificates of deposit that Alma owned in various banks were also placed in the corporation, so these funds could be divided equally among her daughters. Ordinarily this would prompt the issuance of four equal checks to her daughters to distribute the funds. Not so!

After much research and compiling production figures, the daughters decided to use the funds to improve the farm, although mother Alma's "seed" money was not enough to complete the improvements. Financing was arranged. The home farm was especially suited to raising rice because of the abundance of water and the heavy soil that would hold water. However, much needed to be done. The land had to be graded. Some buildings had to be eliminated. Trees that were in the wrong places had to go. Old farm equipment had to be hauled away. Stock watering tanks had to be buried, for it would no longer be a livestock farm. New wells were needed, and some existing wells were in the wrong places. Turbines were installed in the wells. A new system of drainage with new and expanded ditches needed to be built. Dozens of drain culverts had to be strategically installed, for this was a farm carved out of a flood plain. While all of the improvements were going on, improved crop bases were built. All of these things were accomplished over the years from 1990 through 1992. The costs not only included fees for ditches, grading, wells and turbines, but also factored in was the loss of some crops in the process. Following Ralph's dedication to make the farm fertile and productive, chicken litter was added to build up the fertility because much of the top soil had been moved. It was a time for the daughters to tighten their belts. All of the improvements changed the farm, but it is still possible to see evidence of the "old" and the "new." This can best be illustrated by standing near the eight towering grain bins and looking to the southeast to a grove of huge cypress trees. The trees have been gnarled by decades of lightning and storms, but yet stand as sentinels to remind us that the land was once a swamp.

After the improvements were completed, good things began to happen. Crops were better, much better. Each year seemed to improve on the prior year's productivity. Grain bins have been added so there is enough storage capacity for an entire year's crop. This gives the advantages of enabling faster harvests plus marketing control of the harvested crops. None of these things would have been possible without the hard work and thrift of Joe, Mattie, Ralph and Alma and their love of the land. The current owners represent the third generation. Before this decade is over, some of the land will have been family owned for over 100 years.

A farm is a special place to live and grow. The four daughters and their families have fond memories of walking through the gumbo mud that squished between their toes; of catching crawdads in the ditch; of riding the pony and bikes; of playing in the barns; of pilfering in the attic; and of catching gold fish in the stock watering tanks. What a heritage! They also can feel more secure in that the farm will contribute to their financial well-being in the future. What a gift! Since a corporation can have an infinite lifetime, it is possible, even probable, that future generations that haven't been born yet, will benefit from the family corporation farm. What a legacy!

Dear Mattie & Joe Heath, and Alma & Ralph Hubbard,

Throughout your lives, which included the Great Depression, World War II, and the Korean and Vietnam Wars, your hard work and sacrifice have produced a lasting endowment that keeps giving. I salute you.

–Bill Stewart, August 2008

Joseph and Mattie Heath
c.1955

Alma and Ralph Hubbard
c.1960

Farm Heirs at the
Golden Anniversary Celebration
for Martha Jean & Bill Stewart

Mexican Cruise – June 16, 2007
Front Row: Beth Ehrhardt, Luke, Zora, & Eleanor Stewart-Jones, Bill Stewart   *Second Row*: Teresa Stewart, Jeanie Stewart
Third Row: Justin Ehrhardt, Angel Stewart, Dan Bourzikas
Fourth row: Sheila Stewart, Alice Bourzikas
Fifth Row: Don Stewart, Ruth Richardson

Appendix

I. Farm Plans:
        1931 Ralph Hubbard
        1933 Ralph Hubbard
        1946 Ralph Hubbard
        1998 Bill Stewart

II. Obituaries:
        Joseph M. Heath
        Mattie E. Heath
        Ralph C. Hubbard
        Alma E. Hubbard

III. A Paternal History of Ralph Hubbard
        by his brother David Hubbard

IV. Autobiography by Ruth Hubbard,
        Another Voice in the Hubbard Family

V. Hubbard Remembrances for a Golden Wedding Celebration

VI. The Flood of 1937: One of Alma's Published Articles

VII. Love Letters:
        Ralph to Alma
        Alma to Ralph

VIII. Poems: Three Generations of Farm Poems:
        A Mother's Prayer or Song
            by Alma Hubbard
        In the Shade of an Old Oak Tree
            by Ruth Hubbard Richardson
        The Overhome
            by Teresa Stewart

IX. Riding the First Combine on the Farm – Pondering Changes

The first land purchases were made on December 31, 1908 and contained an aggregate of 479.50 acres. Additional acreage was added to the original purchase to make a total of 943.44 acres.

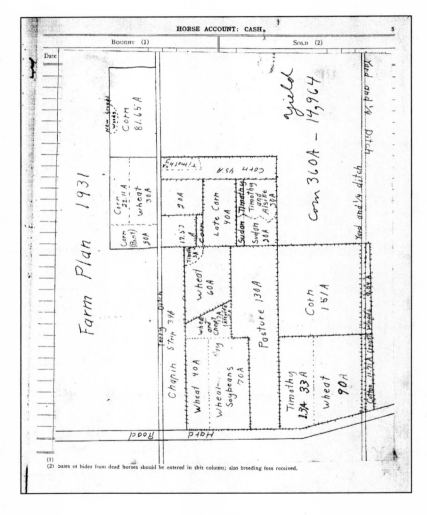

Farm Plan 1931
by Ralph Hubbard

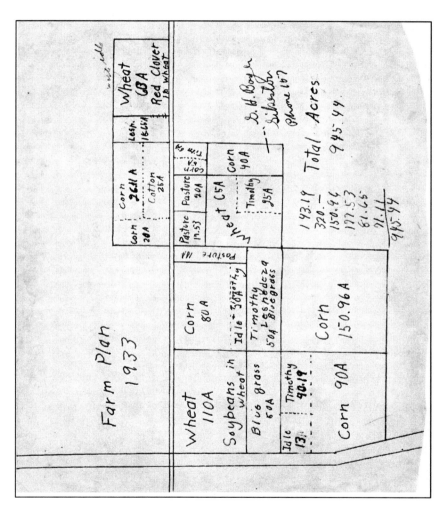

Farm Plan 1933
by Ralph Hubbard

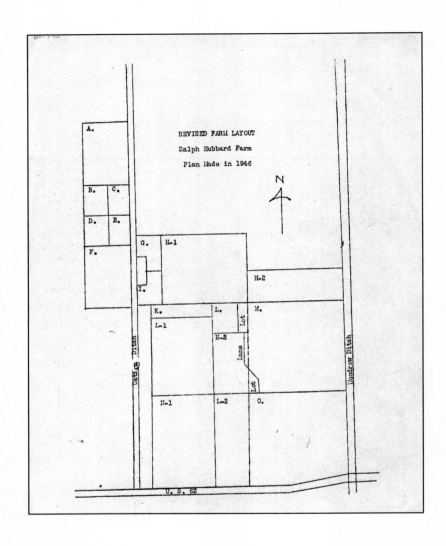

Farm Plan 1946
by Ralph Hubbard

ROTATIONS

	Field Letter	Acres	Planned 1948 Crops	Actual 1948 Crops
1. COTTON ROTATION:	A	55	Cotton (Vetch)	Ogden Soys
	F	55	Soybeans (Ogden)	Vetch under Ogden Soys
2. SPRING PIG ROTATION:	B	12	Corn	Alfalfa (1 yr early)
	C	12	Alfalfa (Tim)	Alfalfa
	D	12	Alfalfa	Alfalfa
	E	12	Alfalfa	Ladino
3. FALL PIG ROTATION:	G	10	Mixed Clover	Mixed Clover
	I	10	Mixed Clover	Mixed Clover
	K	12	Corn	Corn
	L	10	Mixed Clover	Ladino
4. PASTURE ROTATION:	O	143	Wheat	Wheat-Lesp.
	L-1 & 2	141	Pasture	Pasture 65 A. Pasture 46 A. S-100 Soys 30 A.
	M	150	Ogden Soys	Ogden Soys-100A. Corn 50 A.
	H-1 & 2	145	Corn (Soys)	Corn 65 A. Cotton 80 A.
	N-1 & 2	140	Soys (S-100)	N-1 S-100 90 A. Cotton 20 A. N-2 Pasture 30 A.

Crop Plan 1946
by Ralph Hubbard

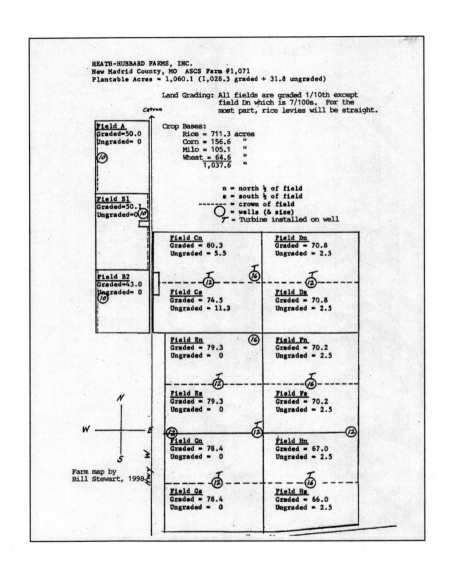

Farm Plan 1998
by Bill Stewart

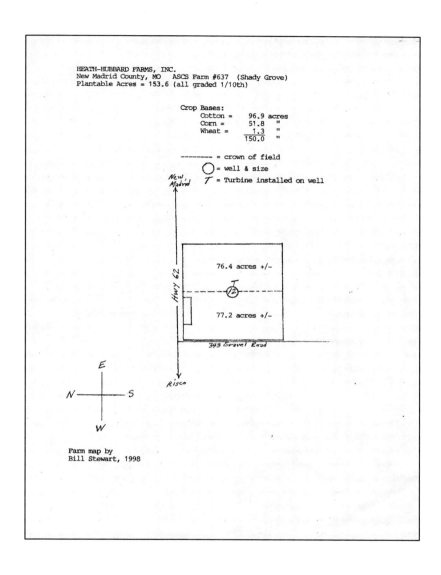

Farm Plan for Shady Grove 1998
by Bill Stewart

Obituaries
as printed in the Local Newspaper

Joseph M. Heath

Joseph M. Heath was born near Greenville in Wayne County, Missouri, on November 7, 1887 and passed away June 30, 1959 at the Delta Community Hospital in Sikeston after a long illness of leukemia. At the time of his death he was 71 years, 7 months, and 23 days old.

Mr. Heath was the eldest son of Christopher C. Heath and Nancy Katherine White Heath. On August 1, 1909, he was married to Mattie Ellen Sweet at Cascade, Missouri. To this union one daughter, Alma Etta, was born. Mr. Heath spent his early life in Wayne County. In 1918 he moved to New Madrid County where he has been engaged in farming for the past 40 years.

In middle age Mr. Heath was converted and baptized in the Christian faith, and this faith sustained him through his long illness.

At the time he moved to the Catron Community, the land was undeveloped. He saw the opportunities and helped with the development of his community. He, with others, was active in the establishment of the consolidated school district of this area, serving on the school board, and his efforts contributed to the first rural transportation of high school pupils in the county.

Mr. Heath leaves to mourn his passing wife, Mattie Ellen Sweet Heath; one daughter, Mrs. Ralph C. Hubbard of Catron; four granddaughters, Mary, Ruth, and Alice Hubbard of Catron, and Mrs. W.D. Stewart of Poplar Bluff; two brothers, Lawrence E. Heath of Flint, Michigan, and Ezra E. Heath of Cowan, Indiana; two sisters, Mrs. Donia Dunn of Muncie, Indiana, and Mrs. Cassie Chatman of Malden; and numerous relatives and friends.

Funeral services were conducted at the Lilbourn Methodist Church on July 2 with the Reverend Robert Hammerla in charge and music by the Methodist choir. Burial was in the Mounds Park Cemetery with Ponder Service. The pall bearers were Homer Baehr, J. O. Wills, William D. Stewart, Oren Ross, Lynual Schuerenberg and J.B. Henson.

Mattie Heath

Mattie Ellen Sweet was born October 22, 1889 at Piedmont, died November 16, 1978, at the age of 89 at the Sells Rest Home in Matthews. She was the daughter of Sterling and Nancy Harrison Sweet.

She married Joseph M. Heath on August 1, 1909. He preceded her in death.

Survivors include one daughter, Alma Heath Hubbard of Catron and four granddaughters, Mary Ellen Jones of the home, Martha Jean Stewart of Poplar Bluff, Alma Ruth Richardson of Portageville and Alice Bourzikas of St. Louis, nine great grandchildren and other relatives.

Funeral services were conducted at the Lilbourn Methodist Church Saturday with the Rev. Bennie Wilsey officiating.

Burial was in Mounds Park Cemetery near Lilbourn with Ponder Funeral Home in charge of arrangements.

Pallbearers were Bill and Don Stewart, the Rev. David Richardson, Dan Bourzikas, Homer Baehr and Dick Twitty.

Ralph C. Hubbard

Ralph Charles Hubbard, 74, of Lilbourn Rural Route, died Wednesday, July 6, 1983, at the Missouri Delta Community Hospital in Sikeston.

He was born on December 28, 1909 at Urbana, Illinois, to the late Fred Clark and Martha Koehn Hubbard. He was a prominent land owner and farmer and an active member of the First United Methodist Church of Lilbourn for years, having served on the Board of Stewards and teacher of the Methodist Men's Sunday School Class and member of the Lilbourn School Board for years. He was always active in community and civic affairs.

On June 4, 1933, he was united in marriage to Alma Heath, who survives. Other survivors include: four daughters, Mary Ellen Jones of Catron, Martha Jean Stewart of Poplar Bluff, Ruth Richardson of Warrenton and Alice Bourzikas of St. Louis; four brothers, Robert K. Hubbard and David Hubbard of Urbana, Joseph E. Hubbard of Peoria, Ill. and Linus O. Hubbard of Chicago, Ill.; and nine grandchildren.

Services were held at 10:30 a.m. Friday, July 8, at the Lilbourn Methodist Church with the Rev. R. L. Hester, officiating.

Burial was in Mounds Park Cemetery with Ponder Funeral Home in charge of arrangements.

Alma E. Hubbard

Alma Etta Hubbard, 79, of Lilbourn died January 25, 1990 at Lucy Lee Hospital in Poplar Bluff. She was a former school teacher and a homemaker.

Mrs. Hubbard was born November 14, 1910, in Coldwater. She was a member of the First Methodist Church and active in United Methodist Women. She was active in extension work. She was a 4-H Club leader for 15 years. She was married to Ralph C. Hubbard, who died July 6, 1983.

Survivors include four daughters, Mary Ellen Jones of Lilbourn, Martha "Jeanie" Stewart of Poplar Bluff, Ruth Richardson of Dexter, and Alice Bourzikas of St. Louis; and nine grandchildren.

Visitation will be at 7 p.m. Saturday at Ponder Funeral Home in Lilbourn. Services are set for 2 p.m. Sunday at the Methodist Church in Lilbourn with the Rev. Belva Rick officiating. Burial will be in Mounds Cemetery.

Memorials may be made to The First Methodist Church in Lilbourn.

Mounds Park Cemetery, New Madrid County, Missouri

A Paternal History of Ralph Hubbard
by his brother, David

When our twelve-greats grandfather was three years old, he was brought to America by his parents on a sailing ship. It was in 1633. He was bonded to the Massachusetts Bay Colony until he was thirty-one years of age to pay for his passage. The father was bonded to the Colony for six years to pay for his passage. The colony had been started in 1630, just three years before they came to America.

Our mother was born in Germany and came to America as a small child. For many years none of the family would tell about the trip. When mother was in her seventies, she said they would not talk of the trip over as it was so terrible. She said about 180 men, women, and children were carried in the hold of the ship with no bunks, partitions, plumbing, or other facilities. All became very sick. Her ten-year-old sister was among the third of the passengers who died on the trip. After over seven weeks on the ocean, they reached New York. To put this terrible experience out of their minds they simply would not talk about it.

Our grandfather, Linus G. Hubbard, walked from Vermont to Illinois after serving in the 16th Vermont Regiment of the Civil War. Visiting a family in northern Illinois, also from Vermont, he found a wife, Helen Stanard. He brought her with him to Urbana. He first bought 80 acres in Tolono Township, but it was low and swampy. He had paid $12 per acre for it and was able to re-sell it at the same price. He then bought eighty acres on the south of Race Street. It was quite well drained and suitable for crops. He borrowed the money from his father's bank in Vermont at 12% interest. He built a house in which all my brothers and a sister were born.

Fred and Martha Hubbard, in 1905, married, bought the 80 acre farm between Race, Lincoln, Florida, and Michigan Streets and built the house shown above at 203 W. Michigan Avenue in Urbana Illinois. In 1914, the barn burned, with some damage to the dwelling. When it was re-built at the time, a second floor was added. Martha C. Hubbard is seated on the front porch. All of Fred & Martha's children were born in this house.

Father had 20 acres of the farm in strawberries. This was before refrigeration, and people who had little or no fruit all winter would come from all over Champaign and Urbana to help pick the crop.

Grocery stores were far different then. Small grocery stores were scattered all over Urbana. One did not go in the store and help himself to items and bring them to a cashier. Most groceries were in bulk, and the grocer would fill cartons from the barrel—peanut butter, sour-kraut, etc. The butcher would cut off the meat the way you asked. Mom would phone the grocer and give him her order. That afternoon it would be delivered by horse and wagon. Bananas would be cut off the stalk when you bought one. Gasoline and kerosene were pumped for you at the grocers!

In 1929 the stock market broke, and we were into the big depression. Banks were closing all over the country. Urbana's banks all closed. The Urbana banking Company went belly up first, followed by the old State Bank. Congress then closed all the rest of the banks until they could pass rigid examination. The First National Bank never reopened. With no banks open, there was very little money anywhere. I recall our hired man bringing back the check my mother had written him as he couldn't cash it anywhere. University employees could not even cash their pay checks at the grocery stores! Promptly, our Mayor, Reginald C. Harmon, suggested Urbana Money. On the sample I have, you can note it was issued by the local Association of Commerce and was only good in stores whose owners were members. Later, when banks reopened, Urbana Money could be redeemed. It was at this time that Ralph went to Missouri to save the farm that was mortgaged against the land in Urbana.

When automobiles were first coming into production, almost any machine shop tried to build them. The News-Gazette of January 12, 1991, lists over 150 different autos made in Illinois alone. The Model T Ford provided my first experience with driving and learning all about autos. The "T" was a touring car and in stormy weather one had to raise the roof and add the side curtains to keep out the weather. Heavy clothing and blankets were required as it didn't have a heater. The windshield wiper was operated by hand with a small crank. A tire would not last much over 1000 miles. They were held on the rims with lugs as the drop center rim had not yet been invented. Some autos had solid rubber tires, and they rode quite rough. The Model T had three pedals. Holding the right one down was low gear. Letting it up and pushing the hand brake lever forward gave one high gear. The car had only two forward speeds. The center pedal was reverse and left pedal the brake. Top speed was about 25 m.p.h. For many years, within my memory, top speed permitted on State Highways was only 25 m.p.h. Then they let it go to 35 m.p.h., 45 m.p.h. and a jump to 65 m.p.h. The early autos would fly apart if one sustained too high a speed. There were two hand levers on the steering column. One was the throttle controlling the engine speed. The other was the spark control. It takes about a second for gasoline to explode, so the faster it went the engine would backfire and the crank handle would fly onto the wrist of the cranker. My father had his wrist broken cranking a Model T. The car didn't have a brake on any wheel. The brake was on a drum on the drive shaft and thus only the back two wheels were slowed. A driver's license was not required.

Autobiography by Ruth Hubbard
Another Voice in the Hubbard Family

As a small child I spent hours pumping members of my family for a story, any story, but one of my favorites was the story of my birth. I loved to hear Jeanie tell it to me best, and I can still remember how she would say, "You know what, Ruthie? I can remember when you were born. Mother sent Mary and me to stay all night with Grandmother, and the next morning when we woke up Grandmother asked us if we would like to have a little sister. We said we would, and then she told us that we had one. We didn't believe her at first, but when we went home, there you were!"

Of all the people I have come into contact with, my parents have had the most definite and direct effect on my life. Daddy is a farmer, and if any person has been called to that vocation, it is he. He is sublimely happy to rise at four o'clock, work hard all day, and come home at night covered with dirt and dead tired. He was raised in the city but his father was a dairy farmer, so he learned early about farm life. We all chuckle when we remember that as a boy he ordered guinea pigs C.O.D. without telling anyone because he wanted to raise them as well as rabbits for the Agriculture Experiment Station. He has a case of mounted butterflies which he collected as a boy and as an agriculture graduate. Mother attended Cape State and taught school before she married. She still teaches, although not professionally, and the list of young people who have been in her Sunday school classes and 4-H Clubs would be quite long. We girls have received the most benefit from her instructions, however, since she has made sure that we learn, not only our school lessons, but lessons in practical everyday living also. No woman is a successful mother unless she practices psychology, and in that field my mother is quite efficient.

My babyhood was quite without incident although I did contract a severe case of pneumonia. My parents have always insisted I was a good baby although I have always entertained a vague suspicion that they were comparing me with my older sisters, Mary and Jeanie, who were premature twins and therefore, quite unhealthy. When I was a year old I could say thirty-six words, but I did not walk until I was eighteen months old. I still talk quite a lot and my muscular coordination has not developed spectacularly, so I am inclined to believe that early developments in a child may be indicative of the course his life will take.

Many of my early memories are concerned with the twins. I followed them constantly I believe, but they were always willing to take me along. We used to build a big play house under a tree in the back yard and spend hours there consuming mud-pie dinners, rocking our babies, and keeping our house, which was always one of my favorite games. Sometimes I could hardly keep up with them and their older friends, and sometimes I did not understand when the twins were teasing me. They once told me that I should put my chewing gum on my nose and save it when I ate, so I disgraced the whole family by walking down Main Street in Lilbourn cheerfully eating a sucker, while a blob of gum perched jauntily on my four-year-old nose. Some people are under the impression that it is hard to be a little sister, but they are absolutely wrong. Some of the nicest people in this world are big sisters, and while they may tease and pass on hand-me-down clothes, they will also pet and comfort when things go wrong.

As a little girl, I developed the habit of running up the path to see Grandmother and Grandfather. Only a small pasture separated our house from Mother's parents and their home would have drawn any child, for they are really story-book grandparents. Grandmother can bake cookies and reminisce as well as any grandmother, but she has always been able to also talk about the modern things too. She tries every new product on the market, and I have often had her tell me to look up words in the dictionary, for she certainly does not believe in ignorance if it can be corrected. She is one of the people I admire most, for she is truly a self-educated woman since as an orphan she had little chance for formal schooling. My grandfather, too, has always been willing to do anything for us children, since we are the only grandchildren. He has always saved pennies and chewing gum for the youngest and will be quite disappointed when my little sister Alice outgrows that stage.

I was six years old before I had to make any important adjustments, and that year I not only started to school, but I also became a big sister. I can still remember the disappointment I felt when I saw my little sister for the first time. I had heard people say that the baby was a fine, big girl, and I picture an enchanting, dimpled, curly-haired creature, who would say "Dada" and sit up and play with me. I was quite disgusted when I saw the tiny red bundle of flesh that was Alice, and I remember telling one of my friends that I did not see why everybody thought she was so great. She eventually proved to be all I had expected although she does not possess the dimples and curls. Nobody can take the place of a baby sister, for although I feel the fiercest admiration for my older sisters, the little one seems somehow to be my responsibility, and I want to protect her from anything that could hurt her.

I attended a one room country school for two and one-half years, and I am very glad that I did. Consolidated schools are undoubtedly efficient and much is to be said for them, but in a small school one receives more personal attention. The transition between the grades is easier, for even first graders can listen to the classes of the older students, thus quite painlessly picking up some knowledge. I think I practiced more natural courtesy while attending a small school than I have ever practiced since.

On the playground we always waited for others to offer to let us use their toys instead of asking for them, and I was quite shocked when I started attending a consolidated school, to hear someone demand the use of a ball or bat. I also remember that I was horrified to learn that the fifth graders had boy friends and were contemplating the use of lipstick, for in the country school the little girls did not even play with the boys, and lipstick was never mentioned.

There is only one thing I regret about going to the country school, and that is the fact that I was double promoted. The first year Miss Ethel was our teacher, and she was a prime example of one of the reasons for consolidation, for Miss Ethel failed to teach us. Mother had taught me to read when I was five, so I was able to do second grade work the next year, but the other members of the class had to redo their first year; consequently, I was able to cover the second grade course at my own speed, and at the end of the first semester I was promoted to the third. This move cause me no trouble with my lessons for during my fourth grade year our school was consolidated, and when I went to the town school I was ahead of my class rather than behind; however, I was not and never have been ahead socially. I was always a year behind—a year too young to wear lipstick, a year too young to date, and so on. I think it would have been better if I had stayed in my own class.

I attended the same consolidated school for eight years and during the whole time I rode a bus approximately eighteen miles a day. This was a rather time consuming process, but it afforded an excellent opportunity to learn of human life, for on a school bus one finds a cross section of human life, junior grade youth camp. I have yet to make the major decisions in my life, but I am not afraid of them. I am looking forward to the rest of my life, for life has been good.

—1956

Hubbard Family Remembrances
Thoughts for a Golden Anniversary

A Summary of the thoughts prepared by Martha Jean (Hubbard) Stewart for the Golden Wedding Celebration of David and Frances Hubbard, June 21, 1987, at Urbana, Illinois.

Family history can be an exciting adventure, but truthfully in preparing for this occasion, I have delved through more Hubbard history, genealogy, letters, and articles, and know more about Hubbard history than I ever planned to learn. Uncle David may rightly be called the historian for our family, for it is he who visits, keeps in touch, and records the latest Hubbard "transactions."

In order to attempt to keep this account short and sweet, I should like to outline Hubbard history for you by using the poetic device of synesthesia, a device that mixes sensory experiences. A musician, for example, will hear a color for a sound; or a color may evoke a smell. I should like to use sound to paint a picture of the Hubbard family, for indeed, when I think of the Hubbard family, I do not so much visualize faces as I hear sounds—the sounds of a busy, active family of five boys who were born, grew up, and attended Urbana schools including the University. It was a family of five boys who inherited a more than normal share of ideas, creativity, and energy.

The first sound I associate with the family is that of laughter—chuckles, guffaws, and the hearty laughter of deep baritone voices. It was a family who loved playing with words and embellishing stories. Uncle David's puns are still a part of that word play, and, fortunately or unfortunately, Uncle David has been the brunt of most of the story telling I have heard, for he was the little guy five years younger than my father Ralph. Perhaps my favorite story is an account of little David's first day of school. It seems that my father was commissioned to take Brother David to school and deliver him to the first grade teacher. My father dutifully set forth with David in tow and followed the instructions. The only problem that arose was the fact that the first grade teacher had moved up to teach second grade, so David started school in second grade, and, according to the version of the story I got, stayed in second grade two weeks until the teacher noted that David was smaller that the other children, and when she started checking, realized that he belonged in first grade. However, I am a teacher, and I have my own suppositions. I am sure that the teacher after having three other lively, inventive Hubbard boys started considering what she could do with this small Hubbard boy who could already read and had been trained in the arts of survival by three older brothers. She was delighted to move him out of her room.

Perhaps, however, the most oft repeated story about Uncle David was one that featured his hiking prowess. It seems that when little David was four years old, he left home on a hike across town. That might not have been so bad, for many four-year olds wander away from home and frighten their parents, but David took along his little two-year old brother, Joe. I always visualized David taking Joe's hand and marching him across town. How do you make a two-year old walk that far? Later, in another account, I was told that he took him in a little wagon. I'm not sure that is any better. When the troubled parents started patrolling the streets in search of the two little guys, they encountered them two miles from home. Totally nonplussed by the whole incident, David greeted his father blithely with, "Why, hi, Pop."

Back to our palette of sounds for painting th~~ ~~ portrait. Certainly in the clatter of sounds that su~~ ~~ family there was a continuous blur of game-playing ~~ ~~ the clang of horseshoes, the crack of a croquet ball, the th~~ ~~ of a ping-pong ball. Even a chess game was a noisy affair. No one but Hubbards could spend an hour boisterously discussing a ten-minute chess game. But the sound that intrigued me most through the years was the splash of water in a swimming pool, a swimming pool that was built in the garage of the Hubbard family home on Oregon Street. Oh, ingenious I thought. Now I realize it was not ingenuity but a survival ploy to survive living with five boys. I think I would have built one in my garage, another in the back yard and perhaps one in the side yard if I had had five boys.

The sound that permeated all activities, however, was that of music. We experienced the same love for music in church this morning as about forty Hubbards worshipped and sang songs that Hubbards have sung in the Baptist Church of downtown Urbana for three generations, songs that proclaim a staunch Baptist and Christian heritage. What really blows my mind is the fact that a music-loving mother would insist that all five boys take piano lessons and that all five learn to play another instrument as well. They did so and proudly marched in various Illini bands or played in various musical groups including a church orchestra and family ensemble with the father on the bass viol and mother on piano. Doesn't that sound romantic? But have you ever considered what it sounded like when five boys practiced at the same time?

These are some of the sounds of my experiences in the Hubbard family, but let me close with a few sounds from Hubbard history of yesteryear, some sounds that may offer some interesting highlights to our family portrait. Some are drawn from the book, *One Thousand Years of Hubbard History*. If we start with the immediate past, we should hear the clump of horses' feet, for the Hubbards came to this area from Vermont and were farmers. Horses were used in farming and to deliver milk for the Hubbard Dairy. In fact, each of the five Hubbard sons took a milk route before school each morning. So, plenty of hoof beats must be daubed into any Hubbard account. We should also hear the squish of gumbo on a farm in Southeast Missouri that Grandfather Hubbard bought as an investment, the farm that I grew up on when my father went to Missouri to salvage the investment during the depression years, the farm that David spent his summers on during his high school years. We should also find German words splashed on the Hubbard canvas, for grandmother Hubbard was born in Flotow, Germany, just outside Berlin. She came to America when she was two years old, grew up in Michigan, came to Urbana where she worked her way through University, earned a degree with a major in mathematics, taught school, and married Fred Clark Hubbard.

Should we choose to highlight our portrait with some unusual tints, we could brush in the blood-curdling cry of King Hubba, a Norseman who plundered and pillaged the shores of England and who contributed his name to the mutant form of Hubbard. We would hear the first cry of a Hubbard baby born in the New World in 1630 in Concord, Massachusetts, to George and Mary Bishop. We could listen to the fiery sermon of Johnathan Edwards who performed the wedding ceremony for one set of our grandparents. We could hear a grandfather fire a rifle in the Revolutionary War, and another Grandfather blast a rifle in the Battle of Gettysburg. For me, perhaps the most intriguing color that seems out of place on the palette of sound is that of the name of a grandmother. Did you know that one grandmother in the Hubbard history was named Submit Hastings? S-U-B-M-I-T! Can any modern day Miss imagine being called "Submit, Submit" by her husband? Enough of Hubbard history!

The Flood of 1937
One of Alma's Published Articles

If it has not been made clear, Alma Hubbard met the fellow members of her Round Robin through a newspaper column. The following is a published column of notable historical significance to the area after the flood of 1937 by Alma Hubbard, signed "Swamper" taken from the St. Louis Daily Livestock Reporter in Hope Needham's column, "The Household Department".

A VISIT FROM "OLD MAN FLOOD"

Dear Householders: "Old Man Flood" and "Old Mr. Flu" have kept me rather busy recently, but I believe I'll call "time out" and visit with you.

The drifts brought by the flood have been cleaned from my front lawn, and at last the curtains are up again so this is more like home. Tulips are peeping up, roses are leafing, men are busy plowing and gardens are planted. Can this be the same flooded land of a few weeks ago?

No, "Old Man River" didn't quite get me, but he was so menacing, and the local flood was so deep that I ran to higher ground. Our location was in an area of danger. Our farm is 12 miles west of New Madrid, Mo., so our safety during the recent flood depended on the strength of the new untried Birdspoint-New Madrid spillway levee. This levee held the Mississippi river after the water was let into the Missouri spillway to relieve the pressure of the Ohio river at Cairo, Ill. We hoped and thought the levee would hold—and it did. We are truly thankful it did hold.

Those in the Ohio valley have my sympathy. When the levee broke at New Madrid in 1927 the water was five feet deep in our home (my mother's home, I was in high school then). From experience I know it is a task to clean after a flood—to say nothing of loss. It is so discouraging to return to a ruined home—

furniture covered by flood water is practically worthless. Even the kitchen range (which you might think could "stand it") had the cementing material soaked from around the oven and was soon found worthless.

An Unkind Winter

The weather was rather unkind to us this winter. In January we had such a heavy ice that trees, telephone poles and electric lines were broken. Then before the telephone and electric service was restored a flood was on us. It rained and rained for days and nights—such rains as we had never had. This southeastern corner of Missouri is a reclaimed drained swamp and it surely looked as if it had reverted to its original state. The water poured over roads and was in most homes around us—our house is built on a high terrace so it lacked two feet getting in it. Suddenly it dropped to freezing; and then it was just a frozen lake for many square miles, with only parts of highways and "ditchdumps" out of the water (and these covered with inches of ice). We thought it was only local and would soon pass, but the water and ice remained and more sleet came. Soon the radio was telling of the flood in Arkansas, just south of us, and of the Ohio flood coming down on us from the north. Then the spillway was flooded and its levee was new. Our few remaining neighbors began to move to higher ground. At this time we became rather anxious about the future safety of our live stock. The ice remained—strong enough to hold the weight of a person, but not strong enough for mules or cattle, and heavy hogs could not walk on it. We waited and worried and realized that we dare not wait for the crest to try our levee, for if the levee failed our live stock would drown before we could move it with conditions as they were.

We had 150 cattle (some purebred), 250 hogs and 20 mules to move. To make the situation more difficult they were marooned on a high place at the center of the farm. It was necessary to cut a road through the ice for a mile. The men worked a day and part of a night cutting and clearing the road of ice. (The nights were really lovely then. Clear sky, full moon, on a land covered with ice. We could scarcely appreciate the beauty, or sleep, for trying to plan a way out.) The men swam the live stock out to the road and drove most of it to higher ground 10 miles away and shipped some to market. They waded ice water up under their arms dragging the large brood sows through the currents. Some they tied by their noses to back of a boat and brought out. They were several days getting the live stock out because of the condition of the roads and the ice, but they succeeded before the crest reached us. They surely worked. They drove live stock, scaffolded wheat and lespedeza seed, piled baled hay, tied wagon beds and bridges. As the levee held the work was unnecessary, but it was too great a risk to wait.

Flood Tragedy

The most distressing tragedy of the flood was the sinking of a barge load of men in the New Madrid spillway. Three men from the village just three miles north of us were drowned—31 lives lost in all. One woman lost her husband and her brother on the barge; another was left with a family of several children—the oldest one 14 years old. The father, a laborer, leaves his family unprovided for. He had little property and could have easily moved to higher ground for safety; instead he (like many others) was working on the levee to protect the property of others. It would seem fair to me that his widow, and others like her, should receive some compensation to help with her family, as would the widow of a soldier. He was fighting to save our country from a foe in the form of flood.

Many amusing things happened as well as tragic. Let me tell the joke on me. I left home, for I was afraid if I waited I would have to bring the twins (age three) out in a boat and expose them to the severe cold. But I was so anxious to return after the crest had passed that I tried to return before the road was in condition and the result was the cars stuck solidly and I finished the trip in a boat paddled with a scoop shovel!

Oh how nice it is to be home again, even if it was greatly mussed! Dad and hubby had stayed at our home all the time and had "batched" with a crew of nine men. How unrestrained we felt at home after having tried to keep the girls quiet in a house of school teachers and older people (although all were very nice to us). But children are noisy, you know! Martha expressed our feeling so well when with a sigh of relief she tells Mary, "You can kie in the musser's house. Couldn't kie at yadies house. Can kie here, Mary."

"Straightening out" the house was just like moving. The men carried down boxes, pictures, rugs, books, etc., out of the attic for an hour, and then remarked laughingly that I must have looked very much like an ant as I tugged so many large boxes up our small attic stair. I was afraid in the rush they wouldn't have time for my furniture.

Enough about the flood! I've talked too long now on the subject. Fortunately our part of the flood was not so serious as in the Ohio valley. However, I can't easily forget the tenseness, the sleepless nights of worry and the hard long hours of exposing work of the men. How one worries when all you possess is in the path of destruction and can hardly be moved! I was in the battle, not on the "firing line," but saw how you who were hit did suffer. My heart goes out to you in sympathy.

I've been re-reading part of "Main Street" (Lewis) and enjoying it—I felt in a "reforming" mood, too.

How's the baby boy, Ex-Steno? You weren't as lucky as my mother's friend. With two boys in college and two in high school she is now proudly rocking her baby boy! What has happened to our robin? Does he fly again? How I enjoy him.

Enjoys Travel Letters

I enjoy our "travel letters." Our excursions are short because of our little girls. We did enjoy a trip to the Big Springs Park at Van Buren, Mo. It is the second largest spring in the world. It was so cool there during last summer's heat.

I also enjoy the letters of parents dealing with childhood problems. But "Future Mother," I believe your ideas are rather "warped." Having your own wee one, I believe would teach you many things. We think we teach them, but, oh, how much we learn as we watch them! Yes, I thought proper environment was so necessary and important (and still give it plenty of credit). But after observing our twin girls' reactions to the same environment and seeing how differently they respond, I'm sure I stressed it too much. If I remember correctly you placed so much stress on proper environment—or just a few children only in homes of plenty. How do we know that the best environment is one of plenty of opportunities? I have learned that what is best for one is not always best for another. It has often been said that "necessity is the mother of invention"—a desire born of need has spurred many people farther than a life with every need satisfied. It is true that the "weeds are out-breeding the wheat" in America, but it seems the "wheat" can never find the convenient time for rearing a family. I didn't mean to sound so critical, "Future Mother," but don't rob yourself of the joy and privilege of motherhood by waiting too long.

Oh, my, where is my bonnet—I know I've stayed too long this time, but I always do stay too long! Hope your dinner isn't late, too, because I felt like visiting today.—Swamper, Mo.

A College Love Letter
From Alma to Ralph

Thursday Eve.

Hello dear,

"In my dream it seems your face is near to me, and is dear to me..." as those words come over the radio they express my thoughts. So often during the day I think of you — and especially at church (when my thoughts should be on other things) I build air castles. But anyway summer term will soon be half over! Hurrah!

I don't know why I'm writing 'cept tis like a visit with you — as nearly as I can visit just now. All the girls are on the poarch chatting except me, and I'm writing you and mom and then retiring.

Next week brings several test, and one instructor tells us he has failed as high as 14%. So guess maybe I'd better review a little.

I have nine pupils and tomorrow I give them a test. Poor dears!

Well I've had my visit so good night dearest. Hoping to see you soon
Alma.

c. 1932

A Birthday Love Letter
From Ralph to Alma

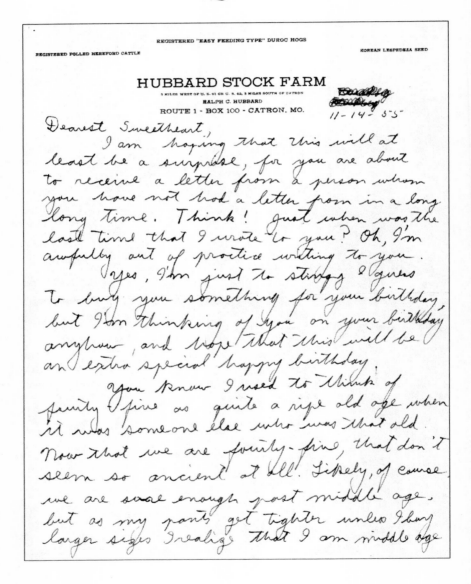

REGISTERED "EASY FEEDING TYPE" DUROC HOGS
REGISTERED POLLED HEREFORD CATTLE
KOREAN LESPEDEZA SEED

HUBBARD STOCK FARM
3 MILES WEST OF U. S. 61 ON U. S. 62, 3 MILES SOUTH OF CATRON
RALPH C. HUBBARD
ROUTE 1 - BOX 100 - CATRON, MO.

11-14-55

Dearest Sweetheart,

I am hoping that this will at least be a surprise, for you are about to receive a letter from a person whom you have not had a letter from in a long long time. Think! Just when was the last time that I wrote to you? Oh, I'm awfully out of practice writing to you.

Yes, I'm just to stingy I guess to buy you something for your birthday, but I'm thinking of you on your birthday anyhow, and hope that this will be an extra special happy birthday.

You know I used to think of fourty-five as quite a ripe old age when it was someone else who was that old. Now that we are fourty-five, that don't seem so ancient at all. Likely, of course, we are sure enough past middle age, but as my pants get tighter unless I buy larger sizes I realize that I am middle age

Side 2 – Ralph to Alma

in the middle as well as in years.

As the years go by, I want you to realize how much I have enjoyed and appreciated the privilege of living with you.

I wish that I could do something extra special today, since today is your birthday, but any how I wish you a very happy birthday!

With Love
Ralph

Three Generations Of Poetry

A Mother's Prayer or Song

Little girl, as I sing to thee,
I'm silently praying that you may succeed
In being the woman I've longed to be.

That your song may ring true—where mine was flat;
That your poems may be formed—while mine is unborn;
That the pictures you paint will interpret God's part,–
Such scenes are not lost to me—but are locked,
For my hands are unskilled in tasks of art,–
That the words of your lips will strengthen rights' cause,
Encourage the weak, inspire the strong;
I tried, but I stammered and my words seemed wrong,
These are the things, I'm praying for thee.

No, I've not failed in all I've tried,
But the things of skill I've been forced to pass by.
My hands can hang a washing white,
Bake sweet cakes, trim true a light,
Bathe a wee babe, make a home gay and bright,
And help my neighbor for friendship sake.
And I pray that you too in these will delight.
Be what I am, and what I've longed to be.
These are the things I want for thee.

— Alma E. Heath
July 1936

In the Shade of an Old Oak Tree

Heaven for me will surely be
A place in the shade of an old oak tree
Where I can play with my sisters three
And our dollies that come for lunch and tea.
The wonder of childhood must eternally be
There in the shade of that old oak tree.

I would like to spend my eternity
Where my mother, her mother, and all before,
And my children, their children and all to be
Can be children together eternally.

The cares of the world will be swept away
In the make-belief world of children at play
Where all play fair, and all may share,
In the beautiful time of eternal care
In a place in the shade of the old oak tree,
An eternal shelter for humanity.

— *Ruth Hubbard Richardson*
1963

The Overhome
(A Memory for My Mother on Her Birthday)

I sweet remember for you the trips to the overhome,
How three children were ferried away in a car that tasted like blue vinyl
And smelled like a plastic trash holder used to catch warm remains of child lunch,
Or hold the fantastic furry worms crossing our road.

   How three children danced in a dust devil, sang for the Do Lord to remember,
      And slept warm away the long pilgrimage to the overhome,
   Until the hollow wooded bumps of the here-we-are bridge would rouse us,
      Trembling, to pass safely over the stack of thick black boards
   Separating us from the deep canyon where live the pinching creatures
      Who liked the fat from grandmother's bacon lowered on a long string.

How three children emerged from the car,
     While you emptied the cold remains of child lunch or grateful worms
And were welcomed to the overhome:
    Where ever-kittens lived under the house,
    Loud chickens made bright orange breakfasts,
    Golden fish swam up from the earth into concrete tanks,
    Already-tree-houses grew in the fields,
    Sheds hid treasures of pink plastic snuff pipes,
       And small tin boxes which made granddad grin,
       And where we ate ice cream sandwiches and had gravy whenever we
          wished.

   How three children marveled at leg-heavy dirt that made pottery and pies,
      And at a house that was really two—with an, ah-attic
   Holding the marvelous secrets of the past-away people,
      And the fragile books and toys of when-I-was-a-little girl.

                                        – Teresa Stewart
                                              C. 1984

Riding the First Combine on the Farm
Pondering Changes
by Martha Jean Hubbard Stewart

On a glorious fall day, Dad appeared at the door of the back porch with a summons for me, a twelve-year old. "Martha Jeanie, how would you like to ride the combine and help me get the soybeans out this afternoon? I can't find any of my men to help me on a Saturday. The weather is perfect. The beans are ready, and we won't have to fight mud."

I responded quickly, "Oh, yes, if you think I can."

"Well, it involves riding on the combine while I pull it with a tractor, and you sew up the sacks of beans and switch the funnel on the chute to fill the sacks. I have the combine ready to go." My years of 4-H would pay off. I could sew!

I got my straw hat, and we headed out to the combine in the field behind the house. There stood the shiny two-row combine in a field of dry beans, laden with their prickly pods, just waiting to burst forth with the first touch of a shattering combine. This would be my maiden voyage on a combine. The big piece of machinery fascinated me. Like an experienced field hand, I climbed into the cab area of the combine where I sat on a wooden bench in front of two gunny sacks that hung on a row of hooks on either side of a pipe-like funnel. The funnel could be switched to fill either sack, but had to be done by hand, no automated stuff on this machine. I was the automation. My task was to hang the unfilled sacks and sew the filled ones. The pre-cut strings hung on a hook beside the sacks with a giant needle already threaded. Dad demonstrated his sewing skills as he wrapped the thread around the corner of the sack and secured it making an ear, then with a whip stitch he sewed the sack closed, making another secure ear on the opposite side. I could do it!

I marveled at the metal slide on the side of the combine, used to slide the filled sacks out of the combine when we reached the end of the field. How efficient! The sacks of beans would be loaded in a pick-up truck at the end of the day. Never did I dream that there would be a day when a self-propelled combine would have its own hopper to collect the beans and by-pass my lowly sewing job. Grain bins would take the place of the gunny sacks.

We were on our way. I funneled and sewed the beans that were cut with the shaking blade and rolled up the elevator of the combine into the sacks. I watched the debris that dusted out from under the combine while only the golden beans poured into the sacks. We were covered with layers of dust and grime, but so happy to see the bean sacks pile up at the end of the field.

Years later when riding a train from Shanghai to Hangzhou in China in 1998, I watched Chinese workers carrying baskets of rice on a pole across their shoulders along narrow paths between the rice paddies to a threshing floor. Then I watched as workmen tossed the grain by hand up in the air and beat it with a flail, a device that look like a wide band of black rubber attached to a wooden handle. How primitive, I thought–in 1998. Today we ride a self-propelled air-conditioned combine, listen to music on the radio, have computers to help assess tasks, and push a button to tell us how much grain is cut from the rows on one trip through the field.

I am seventy-five years old and can remember when I watched a man hold a single plow head in the ground while a team of mules pulled the blade through the compacted soil. I remember when I helped harvest soybeans in gunny sacks, and when I tramped down cotton in a field wagon.

What changes in farming practices will the next generation explore?